MEMOS FROM EGYPT

JOINT U.S.-EGYPTIAN STUDY GROUP ON BUILDING MATERIALS AND BUILDING TECHNOLOGY.

Cahners Books

Library of Congress Catalog Card Number: 75-13629

International Standard Book Number: 0-8436-0157-4

Cahners Books, a Division of Cahners Publishing Company, Inc.
89 Franklin Street, Boston, Massachusetts 02110

Printed in U.S.A.

Foreword

When I first began to read *Memos from Egypt,* I was turned off by the beginning of the book, "This Is When It All Started," which implies that the book is a private memorandum and travel diary. But my uneasiness gave way to fascination with such personal aspects as hotels, airplane connections and the inconveniences of travel; this description also provides excellent criticism of airport design.

Bill Caudill is one of the United States' foremost architects. He has written a book that to some readers may at first seem too intimate for universal meaning, yet it is this very informality and intimacy that establishes a frame of reference for his evaluation of the Egyptian scene. As such his comments are invaluable—what you have here is good technical data concerning building, the state of the building industry and the trades in Egypt—told in a warm, personal way by one of America's experts.

The information contained in *Memos from Egypt* is first-hand, objective and incisive. It is an excellent travel handbook for architects, engineers, contractors and builders who intend to do business in Egypt. Even for those who don't have any immediate plans for work in Egypt, the book will make interesting reading because of its lively and informative descriptions of an exciting trip.

Forrest Wilson, A.I.A.
Chairman, Department of Architecture and Planning, The Catholic University of America

Thanks to Sandy Gumienny, Jean Lucas, Jeffry Corbin, Amy Broussard and Aleen Caudill for helping me put this Diary to bed. W.W.C.

Bill Caudill left Houston for Cairo, Egypt, on 15 January 75 to serve as a member of a Joint U.S.-Egyptian Task Force to study building materials and building technology.

Each day Bill sent his impressions back to our offices. We read the memos as they drifted in and were fascinated not only with the unequaled architectural heritage of Egypt which Bill described so vividly, but also with his thoughtful insight into the tremendous building needs of the country. We thought others might like to read about the trip, especially our friends.

The U.S. delegation is developing an official report for the State Department. Later there will be a public version available for distribution by the Building Research Division of the National Bureau of Standards.

We hope you find Bill's impressions as stimulating as we did.

Chester L. Lucas

Chester L. Lucas, President, CRS Design Associates International, Inc.

EGYPT
75

THIS IS WHEN IT ALL STARTED.

16 January 75 - Enroute from Houston to Dallas

1. This is the day! First to Dallas. Then to Washington. Then to London. And finally to Cairo.

2. Got to go over my checklist:
   Ticket
   Money
   Passport
   Lomotil (the international pill)
   Halazone (the water pill)
   Granola and safflower oil (to lower my cholesterol)
   Aleen

Better not forget the last one. ☺

MARGINAL NOTES AND SKETCHES THAT FOLLOW ARE THE PRIMITIVE WORKS OF THE AUTHOR.
THE AUTHOR

3. When Jim first called to invite me to go with him and his group to Egypt, I quickly responded, "I'd love to, but I've got a domestic problem." "What?" he asked. "I was

EGYPT (UNITED ARAB REPUBLIC) LIES AT THE SOUTHWEST CORNER OF THE **MEDITERRANEAN SEA**, WHICH HISTORICALLY LINKED IT TO EUROPE.

ASIA WAS REACHED BY SHIPPING ROUTES THROUGH THE **RED SEA**.

married two months ago. I couldn't possibly go without taking Aleen. She's my cure for senility." "Take her," said Jim. He didn't volunteer for Uncle to pay her expenses, however.

4. Who's Jim? Jim is James L. Haecker, Industrial Liaison, Center for Building Technology, National Bureau of Standards.

5. What trip? I go to Egypt as a member of the Joint United States-Egyptian Study Group on Building Materials and Building Technology. Jim is the appointed coordinator of the U.S. task force.

6. Members of our task force will be: Phillip D. Bush, VP, Kaiser Engineers, Oakland, California; Professor Neils Thompson, C.E., University of Texas, Austin, Texas; William C. Cullen, Asst. Chief, Office of Housing Technology, Center for Building Technology, National Bureau of Standards; James L. Haecker, Industrial Liaison, Center for Building Technology, National Bureau of Standards; and The World's Greatest Architect.

7. Why was I invited? Why do you ask?

8. Getting to Cairo is no picnic. Aleen and I were scheduled to leave this morning at 10 a.m. on Braniff; get to Dallas at 10:50; take off on American at 1 p.m; and arrive in Washington at 4:50. We'll meet Jim Haecker at Dulles airport in Washington and, with Bill Cullen, will head for London on Pan Am at 7:45 p.m. We get to London tomorrow at 7:45 a.m. Whew! We're going to feel like we've been to Cairo and back before we even arrive.

9. Jim Haecker had originally scheduled us on an 11 a.m. plane to Dallas/Fort Worth, leaving us 1 hour and 10 minutes to transfer from Braniff to American. I raised hell. Any ransfer at that airport takes at least two ours. Reminded Sandy Gumienny, my secretary, of my last experience in "the world's largest airport" from which Aleen and I barely escaped. On that occasion, after two hours of fighting nutty computers, obstinate turn-table gates, stalled trains and a malfunctioning cafeteria, I concluded that going through the Dallas/Fort Worth airport is a career.

10. So Sandy changed our reservations from an 11 a.m. Houston departure to 10 a.m. We did

OVERLAND ROUTES ACROSS THE ISTHMUS OF **SUEZ** AND **SINAI** PENINSULA JOIN ASIA MINOR AND INNER ASIA. THE **NILE** AND ITS TRIBUTARIES WERE PASSAGES INTO EAST-CENTRAL **AFRICA**.

EGYPT DURING THE DYNASTIC ERA WAS A RICH MARKET FOR BOTH NECESSITIES AND LUXURIES.

AFTER THE SPREAD OF **ISLAM** (650 A.D.) EGYPT CONTINUED AS A MAIN WORLD TRADE CENTER.

CAIRO WAS BUILT IN 969 A.D. MUCH OF ITS BUILDING MATERIALS CAME FROM THE WRECK-AGE OF ANCIENT **MEM-PHIS**.

POPULATION OF **CAIRO** TODAY IS ABOUT SIX MILLION.

even better than that. Sandy drove us to the airport so fast (and in the fog too) that we made the 9 a.m. plane. I'm writing this on that plane and thinking: Surely we (Aleen, baggage and I) will be able to make a successful hook-up with American to Dulles Airport in Washington. Wait until the next exciting episode to see if we escape the crushing clutches of D/FW.

11. We'll be staying at the Skyway Hotel in London, which I understand, is close to Heathrow Airport but not convenient to anything else. Friday we'll try to sleep. Friday night we'll have dinner with old friends, Sir Antony and Lady Part.

12. On Sunday, 19 January, Jim Haecker, Bill Cullen, Aleen and I leave London on Egypt Air at 2:15 p.m. We land in Cairo at 10 p.m. where I'm told we'll be given a small royal welcome. We bed down at a place called the Mena House and get set for a week of hard work.

13. On Sunday, 2 February, Aleen and I (Jim will return a week earlier) head back to London. Will stay at the Skyway Hotel again. After some more visiting and sightseeing, we head for home

on Wednesday, 5 February, at 10:50 a.m. and at 7:19 p.m., Wednesday evening, a couple of exhausted travelers come home to Houston. It's going to be a busy ~~two~~ 3 weeks, but I know it'll be great.

14. D/FW airport coming up. Got our fingers, toes, and even our eyes crossed in hope.

Bad sketch
of the author
by the author.
Editor

One Hour Later -- Dallas/Fort Worth Airport

1. We made it from Braniff terminal to American. No trouble whatsoever. The little train got us here in good shape. Its computer was working. The tape recorders were synchronized perfectly with the stops. Now if the baggage gets on the same plane we do, everything will be great. The last time I transferred in this airport, I made the connection, but I had to run like hell. My baggage couldn't run that fast. It arrived in Houston the next day.

2. My other experience in the D/FW airport was limited to the Braniff terminal. What a contrast.

Braniff's spaces intimidate the hell out of me. To my eyes, they're nearly obscene, inhuman, vulgar, superfluous forms; over-subdivided; too many materials. It bugs me; so cluttered and confining. Aleen says Braniff's terminal reminds her of stuffed elephants in a playpen with no room for the kid. It's a shame too, because years ago Braniff was proud of its design -- interior and exterior -- its airplanes, stewardess uniforms, service vehicles and terminals. The terminal is a bummer now -- overdone and in poor taste. And very expensive.

"LESS IS MORE"

3. The American terminal where I'm writing this is something else. Elegant though restrained. So simple. No visual confusion. And it works better -- better form, better function, and at less cost.

4. Called Channel 13 here in Dallas and talked with my feminist daughter Susan. She was doing final editing on a film about abortion. Naturally Susan is interested in what and how women are doing in Egypt.

5. This side of the terminal isn't the epitome of efficiency. Tried three phones before I

found one to work. Departure boards are inoperative. Plane about ready to board but Aleen and I still don't know which gate.

16 January 75 - Enroute from Dallas to Washington

1. We made it! We are on American 438.

2. I suppose the real reason Jim Haecker and his boss, Dr. James Wright, Deputy Director, Institute for Applied Technology, National Bureau of Standards, invited me to serve on their team is that the three of us have long been associated professionally -- dedicated to advancing building technology.

3. I first knew Jim Haecker when he was a young researcher at the University of Michigan. At the time he was delving into the use of plastics as a building material. I was Director of the School of Architecture, Rice University. One of my young professors, Charles Thomsen (now President of CM Associates, Inc., the international construction management firm and sister

organization to Caudill Rowlett Scott), had heard of Haecker's experiments and talked me into bringing him to Rice to teach our students about plastics and how to put plastics to work. I thought it was a bum idea, but reluctantly I gave in to Thomsen. As it turned out, our students were turned on. Bringing Haecker to Rice was a damn good idea. In fact, there hadn't been so much excitement among the students since Charles Schorre, another young prof of mine, brought in a belly dancer to serve as a model for a nude dancing among the trees.

4. Jim later served the American Institute of Architects in a full-time capacity in research before he moved on to the National Bureau of Standards where he has helped bring the profession and the government closer to a true partnership. He's a highly competent research architect. And he is greatly respected by the CRS team who brought him in as a consultant when we were writing Bucket of Oil -- The Humanistic Approach to Building Design for Energy Conservation, (authored by Tom Bullock, Frank Lawyer and Old-What's-His-Name.)

ABOUT TRIP AN INSERT

After Jim Haecker invited me to go to Egypt with his group, I received some information spelling out what this trip was all about. It seems in June, 1974 Henry Kissinger and the Egyptian government set up a cooperative U.S.-Egyptian study group for technology research and development. Hopefully exchanges and joint research in the scientific and technical fields will be of mutual benefit to the two countries.

According to the agreement, "the governments will encourage and facilitate the development of direct contacts and cooperation between governmental agencies, universities, research centers and other institutions and firms of the two countries, and will make arrangements between them for carrying out mutually-agreed-upon cooperative activities."

The goal of the program is to intensify cooperation between the scientists and technologists of the two countries by providing them with additional opportunities for the solution of problems of mutual interest. Closer cooperation between the scientists and technologists of the two nations should advance the state of science and raise the level of technology in both countries. And getting together should strengthen the bonds of friendship between the peoples of the two countries.

Included in the information package was a report of the second meeting of the Joint U.S.-Egyptian Working Group for Technology Research and Development which was held in Cairo on November 3-4, 1974. The report lists the five general areas of study:

A. Repair and maintenance of laboratory equipment
B. Water resources management and treatment, including desalination, in relation to agriculture
C. Building materials and building technology
D. Information systems matters
E. Lake Nasser water problems

Our study covers only Section C, Building Materials and Building Technology. In essence we are a sub-committee of the larger committee. What we do will need its approval.

5. The pilot just told us to look out the left window and we could see where the Ohio River runs into the Mississippi. He pointed out that the little town down there was Cairo. Got here darn quick. This Cairo, however, doesn't have six million people and its name is pronounced like the syrup.

KARO?

6. The pilot also called our attention to a lake which was created by a hole made during a 1911 earthquake -- 8 on the Richter scale, greater than the one that destroyed San Francisco. Earthquakes in the Midwest? I didn't know that.

7. Dr. James Wright, Jim Haecker's boss, and I were members of a U.S. delegation to Russia for a three-week study relating to industrialization of the building process. Jim Wright was the leader of our group.

That was five years B.D. (before détente). It was quite an experience, but a most rewarding one. Since then I've served on Jim's advisory panel. He continues to exert strong leadership and recently established the Center for Building Technology to serve as the U.S. technological arm in Building Sciences.

8. Jim Wright reached the highest peak of courage during one of his visits to Houston when he permitted me to fly him around Houston in my little aircoupe -- that's the nearest thing to a mechanized kite. He was brave.

9. Oh, how I wish I could take my current mistress to Egypt along with Aleen. My mistress -- Zero Six LUV -- is a 200-horsepower Lake Amphibian, a pusher which can land on both land and water. Since Egypt's agriculture and industry seem to grow in the Nile Valley, what better way of seeing the country than by a Lake Amphibian. She would love the Nile.

10. I haven't been with our task force yet, but I'm particularly anxious to see Phil Bush. He was with us on the Russia study tour, too. Phil, an industrial engineer, made a major contribution in helping our delegation arrive at cost comparisons between the Soviet structures and ours. He's numerical as hell. As for me, I can't add for sour apples.

11. Phil gave me some great news when I talked to him last week. He's bringing his beautiful spouse. I guess she wouldn't let him go without her. I'm delighted about being with Phil again, and Aleen is very pleased that his Marjorie is going.

12. What a slave driver Jim Haecker has been. I thought Jim Wright pushed us during the Russian study on industrialized buildings, but Wright's boy Haecker is something else.

13. Jim had Aleen and me working on getting our passport pictures updated and shots taken even before we knew our itinerary. I'm allergic to and horrified by doctors, nurses and needles. Aleen was just as scared about getting her picture taken. Why is beyond me. She's a thing of beauty.

14. As for the shots, my doctor, Bill Owen, recommended that we take a smallpox vaccination, a typhoid injection and gamma globulin. He didn't have the guts to stick me himself so I was handed over to a nurse. She treated me as if I were a male chauvinist pig -- stuck me with glee, using a dull needle and an overdose of

HERODOTUS (490-425 B.C.) SAID, "EGYPT IS THE GIFT OF THE NILE." SINCE THE TIME OF THE PHARAOHS THE PEOPLE HAVE ATTEMPTED TO CONTROL THE EXCESSES OF THE **NILE RIVER.**

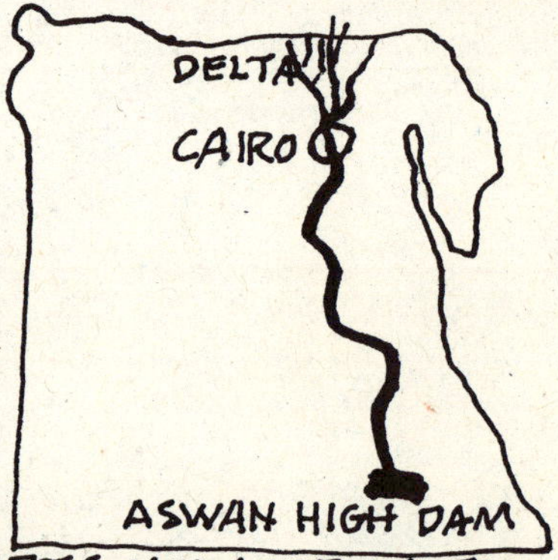

HOPES NOW LIE IN THE **ASWAN HIGH DAM**-- NEW LANDS AVAILABLE FOR AGRICULTURE, OLD LANDS TO YIELD MORE THAN ONE CROP, AND THE POSSIBILITY OF ELECTRIC POWER TREBLING.

stuff in each shot. I hate to hear a grown man cry, particularly when it's me. The nurse was rather put out when I modestly asked to get my gamma globulin shot through my shorts.

15. While getting my shots, I had a flashback to the time Jan Talbot (my loyal secretary then; now she's a big shot in CRS) rode with me in an ambulance to the cardiac ward of a local hospital where we were met by a truly <u>female</u> chauvinist pig, Dr. Jackie Ribelles. Later she told Jan that she needed the help of six nurses to remove my shorts. Understandably, I was reluctant to let this nurse shoot gamma globulin into my bare behind.

16. Apparently the three holes which the nurse put into my delicate bod weren't enough. Later in the day I was stung on my stomach by a wasp, right in our living room. It hurt like hell and itched more than the vaccination which took. Nevertheless, when I saw the nurse later, I told her that, of the two who made holes in me, I preferred the wasp.

17. We're 20 minutes from Washington. Aleen was reading <u>All the President's Men</u>, but she

fell asleep. She's really been working hard this past week putting together all the little pieces so we can make this trip. It was hard enough scheduling my son Bill and her daughter Julie to pick up the Saturday and Sunday papers, but to "agendize" Sallie, our maid, for three weeks to take care of the house and the plants was even worse. Plus buying new clothes and patching old ones, getting hair fixed, getting other things fixed, paying bills, and so on. Aleen didn't want her husband to look like a slob so she bought me a beautiful cashmere sport coat. Putting cashmere on me is like putting tails over jeans. But I always wanted to be a clothes horse. Now I is one.

16 January 75 - Enroute to London

1. We made it into Washington. And with baggage. The luck of the Scotch-Irish.

2. Aleen and I met Jim Haecker and his wife Polly at Dulles. Beautiful couple. Polly had already made one trip to Egypt with Jim, so this was her time to stay at home to take care of

their two kids -- both teenagers. Aleen and I have four between us. Thank God they're grown.

3. Bill Cullen joined us. He'll serve on our task force as the building materials expert. I'm not sure why I'm on the team. Could it be my research years in natural lighting and natural ventilation? When I was involved in this activity at Texas A&M, certain people on the faculty referred to me as the expert on hot air. As my Mom used to say, "No one ever kicks a dead dog."

4. The inevitable happened. Spilled a soft drink all over my pants. One pair down. Two to go.

5. This Pan Am 747 is a biggie. We'll be on it for six hours. We're in economy class, but a nice stewardess let us move into a set of four-in-a-row vacant seats. Beats first class. We now have berths to go beddy-bye.

YOU'LL HEAR MORE ABOUT THIS BOOK →

6. I've been reading an intriguing book -- Architecture for the Poor by Hassan Fathy (The University of Chicago Press, 1973). The book was originally published in 1969 under the title

Gourna: A Tale of Two Villages. The 71-year-old author-architect, Fathy, is most sensitive to people's needs. He has become world-famous for designing attractive, inexpensive homes for the poor, using cheap local materials. A recent CRS design team in Saudi Arabia told us Fathy is highly respected in that country and is used as a consultant on a good number of projects. The son of well-to-do parents, Fathy discovered the poor on his parents' farm. Sounds like an architectural Franklin D. Roosevelt.

Fathy makes his case for the poor. Says that according to the U.N., 800,000,000 peasants -- 1/3 of the earth's population -- are doomed to premature death because of inadequate housing.

7. Fathy speaks of rediscovering the method used by the early Egyptians to build houses using domes and vaults of mud and mud bricks. He accuses architectural professors of neglecting the poor and being blind to sensible, modern use of this very old, most proven material. But the idea of using this old material with new design techniques simply does not interest concrete/steel-minded profs. Fathy advocates the use of mud brick as sole hope for rural

reconstruction. Two reasons for this: a) the mud is there; b) by using it, the peasants can build their own houses.

8. This idea takes me back to 1939 my graduate days at MIT when a classmate and I won a national competition called "The Productive Home Competition." The notion then ~~(this was the time just following the great war and before the depression)~~ was that each family would have five acres on which to raise vegetables, fruit, have chickens, a cow and a pig or two. The competition was to design a home which the family could build themselves. Our winning design put the cows, chickens, automobile and people under one roof -- controversy. But the prize money paid for my tuition, nevertheless. What material did we use? You guessed it. Earth. The climate was not right for adobe or mud brick. We proposed rammed earth blocks developed by experimenters at Oklahoma State University. I'm sure the house or two that O.S.U. people put up is still standing in Stillwater, Oklahoma. Of course the U.S. has thousands of old adobe homes in use today. Earth is a damn good building material if used right.

ALEEN'S AND MY ALMA MATER

9. The CRS team has used earth to a great advantage in some of its most contemporary buildings. The Akron (Ohio) Hall of Performing Arts, which New York Times critic Ada Louise Huxtable praised as one of the best (and she doesn't hand out many roses) has earth berms for energy conservation. So do CRS-designed schools in Colorado and Florida. As I said, earth is a damn good building material. Fathy is right.

EGYPTIAN ARCHITECT **HASSAN FATHY**. IF HE IS NOT A GREAT DESIGNER, HE IS CERTAINLY A BEAUTIFUL WRITER.

10. While I'm writing this, Aleen is comfortably stretched across two seats -- fast asleep. Who needs first class? The food was very good. We paid $2.50 each to see a movie, but I bet we'll never stay awake.

11. Back to Fathy. In spite of the economy (of materials and labor) of mud brick walls, he discovered that the houses themselves were not much cheaper because the timber for roofs ran up the cost. What was needed were mud brick roofs, he thought.

12. Fathy remembered the ancients had built vaults without building wood formwork. About that time, he was asked by Egypt's Royal Society of Agriculture to do some designs incorporating

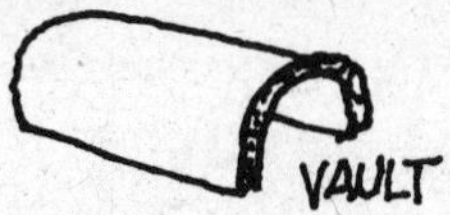

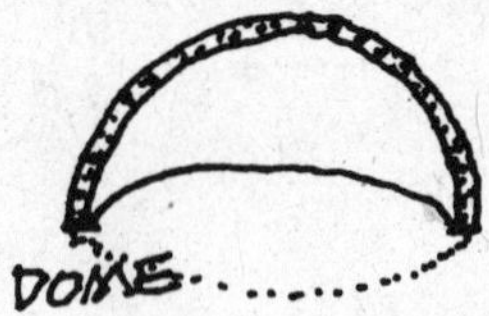

his new ideas in houses. He explained his wishes to the masons to put up the vaults with no formwork. It was a bad trip. The vaults promptly fell down.

13. To make a long story short, after much thought and search he found what he was seeking -- brick vaults without formwork -- in the little village near Aswan.

14. Hassan Fathy puts this discovery so splendidly:

"It was like a vision of architecture before the Fall: before money, industry, greed and snobbery had severed architecture from its true roots in nature."

SEE WHAT I MEAN ABOUT FATHY'S BEAUTIFUL, POETIC WRITING

He concludes that building vaults and domes without wood forms was perfectly familiar to Egyptians in the 12th century. And it's an absolutely necessary economic measure today where wood and steel are scarce.

15. He finally found two Nubian masons who did the trick for him. Fathy describes the event:

"It was so unbelievably simple. They worked rapidly and unconcernedly, with never a thought that what they were doing was quite a remarkable work of engineering."

16. The masons had great intuitiveness for the material without formal education. Nevertheless, they worked according to fundamentals of the physics of materials and statics of structures. The vault was parabolic, conforming to the shape of bending moment diagrams. All bricks were in compression. They would have pulled apart in tension. It makes no difference what kind of materials are used --concrete, wood, steel, mud or butter -- the structural behavior remains the same.

17. Fathy was then able to build his houses for the government. All the houses were roofed with mud. Rooms, corridors, loggias were covered with vaults and domes using a material good in compression and a configuration which caters to compression.

18. So much for stress. My fingers are overstressed now, not to mention my brain.

17 January - Enroute to London ON PAN AM 747, A BIGGIE.

1. The movie, "California Split," started off very well. Aleen never awoke to see it. After 10 minutes, my sound went off. Got my money back and went to bed.

2. Aleen and I did extremely well during the night, each in a fully prone position. The four seats we occupied were not exactly the area of a king-sized bed, but with Aleen's head on the seat of one aisle and my head on the seat of the other, and bods overlapping, we did as well as anyone on the giant bird. About half way over, Aleen rose up and said, "Shift." I knew what she meant. My side was numb too.

3. Three of the overhead lights wouldn't turn off, but we fixed that by putting some cards over them. Then back to the sack.

4. Back to Egyptian architect Hassan Fathy's <u>Architecture for the Poor.</u> He's quite a guy. Real poet. I particularly like this:

"Building is a creative activity in which the decisive moment is the instant of conception, that instant when the spirit takes shape and all the features of the new creation are virtually determined."

5. Fathy expands this thought in just as poetic terms:

"While the characteristics of a living creature are irrevocably settled at the moment of fertilization, the characteristics of a building are determined by the whole complex of decisions made by everyone that has a say in the matter, at every stage of its construction."

You better believe it. When he says construction, I'm sure he means conception as well.

6. CRS has always advocated the team concept and practiced it with varying degrees of success. Call it group dynamics. Call it Gestalt psychology. Call it teamwork. Call it anything you like. Better buildings can result. And they are better because many experts "had a say in the matter." There must be something to the notion of the team or we wouldn't have five pros on this task force.

THE MICHELANGELOS CAN'T COPE WITH TODAY'S COMPLEXITIES. IT TAKES A TEAM. THE **NEW GENIUS** IS THE TEAM.

7. Fathy includes this quote from Dante Alighieri:

"It may be that what we call modern is nothing but what is not worthy of remaining to become old."

Fathy spells this out in his own terms:

"Modernity does not necessarily mean liveliness, and change is not always for the better. On the other hand there are situations that call for innovations. My point is that innovation must be a completely thought-out response to a change in circumstances, and not indulged in for its own sake."

Bull's-eye, Hassan!

8. Fathy also included one of my favorites from Lao-Tze:

"Completeness without completion is useful. Fulfillment without being fulfilled is desirable."

Paul Kennon, LA office, and Norman Hoover, NY office, have both carried this thought to architectural manifestation. They contend that no building is ever finished. Most have to grow. Each has changes in function in its life cycle. They want even old buildings to look young, as if they wanted to grow, wanted to change. These two highly skilled designers and their protégés have carried this notion to an advanced stage, evolving toward an aesthetics of incompleteness.

9. Architect Fathy says, "This artistic crystallization is not a once-and-for-all occurrence, but an act of reaction that has to be perpetually renewed." Beautiful. If words were brick, what a structure he would have.

10. A case in point: My own house. In its 12 years of life, four people lived in it originally, then three, then four, then three again, then one, now two -- with each set living different life styles. Six major architectural changes have taken place during those 12 years.

When function changes, form must respond. To respond successfully, form may require physical changes. Sometimes additions. Sometimes modifications. But at all stages there must be visual "completeness without completion."

17 January - Near London

1. It's one hour prior to landing and they've brought us a continental breakfast. Why so early?

2. Back to Fathy's Architecture for the Poor. He delights in the restraints mud brick forces upon design for large-scale, rural housing. "We are fortunate," he states, "in being compelled to use mud brick . . . poverty forces us to use mud brick and to adopt the vault and dome for roofing. All our buildings must consist of the same element, slightly varied in shape and size, arranged in different combinations, but all to human scale." That's one way of getting visual unity and human scale -- design/building with mud bricks.

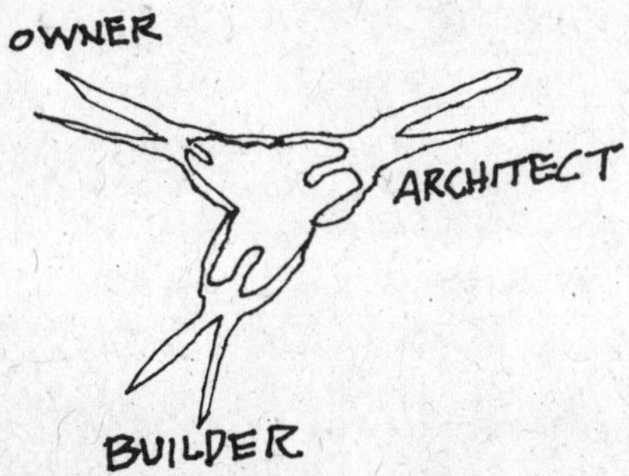

3. This very sensitive Egyptian architect argues for the reestablishment of the "trinity": owner, architect and craftsman. He wants all three on the job, working together. Ideally, he said, if a village were to take three years to build, the designing should go on "for two years and eleven months." He reinforces this notion, "Right up to the last moment I should be learning, modi-

fying and improving my designs."

BULLISH IN HEATHROW

4. Landed in London. No baggage trouble. No customs trouble. They even had a grocery cart available (at no cost) in which to carry our baggage. But we had to walk miles. Did enjoy one stretch of moving sidewalk. Great for carrying heavy bags. Mostly long-haul halls like cattle chutes. I felt I was in the Chicago stockyards. Inhuman. On the other hand, I was impressed with the hand-made signs paging people as we came out of customs. NO SCREAMING OVER LOUD SPEAKERS.

5. Waiting for our baggage, I noticed that Aleen's beddy-bye did the trick. She looked pert. I looked beefed. Why not? I had just come through the stockyards.

6. No trouble getting to and into the Skyway Hotel although it was only 8 o'clock in the morning.

7. Jet lag has begun to set in. Aleen is starting to look like me -- beefed.

17 January 75 - London - Afternoon

1. We're well situated in the Skyway Hotel. Before Aleen and I hit the sack to catch up on our sleeping -- hopefully to alleviate the jet lag -- I called Lady Ella Part. Sir Antony and Lady Part are old friends of ours. He's in the Ministry of Industry. Aleen and I were to have dinner tonight at their flat.

2. Willie, Wallie, John and others who know these beautiful people will be sorry to hear that Antony had what may be an angina (heart) attack early this morning. Ella wisely put him in the hospital for a full check-up. He insisted that Ella come to our hotel to visit us, but I insisted she had no business doing that at this time. I told her perhaps we could get together when we returned to London two weeks from now. I hope all is well with Antony. As I said, they're beautiful people.

3. We're in the throes of withdrawal pains from flying. Mid-afternoon, but hungry. Ate a cold plate and tea in the hotel. Generous portions.

4. Will take a bus to town. To taxi in would be about a $20 fare. Houston and London do have something in common.

5. Aleen asked a porter where in downtown London one could walk safely at night. He retorted, "Madam, this is London. We have no muggings." Read in the paper where crime was so much on the increase police were put on commuter trains and buses. Blind patriotism?

17 January 75 - London - Evening

1. We took a hotel bus (a very good one) to downtown London. About an hour's drive. Passed all of Aleen's favorite shopping places like Harrods, Scotch House, Burberry and a great oriental rug place. Thank God they were all closed.

CRS' HARVARD UNIVERSITY GRADUATE SCHOOL OF EDUCATION IS ENVELOPED IN "HARVARD TWEED" — THE ORIGINAL BRICK USED AT HARVARD.

2. On the way I satisfied my insatiable appetite for brick. Love that brick. Most buildings wear brick ~~clothes.~~ TWEEDS.

3. Was shook up on seeing the plumbing soil stacks stuck to the outside walls like jumbo

flow diagrams painted on huge sheets of textured cardboard. Dramatic, but ugly.

4. Aleen got out her electric curling iron to work on her hair before we went out to eat. Prior to leaving Houston, she bought a converter because she knew the current and voltage were different here. She also had purchased several adaptor plugs. Problem? Her curling iron had a flat-pronged plug which won't fit into these outlets with their round holes. Solved the problem by taking the light bulb (a push-in type) from the bedside lamp and putting in the push-in adaptor plug. I wouldn't touch the switch. Even static electricity scares me. Aleen's braver than I am. She flipped the switch, but I shut my eyes and stopped up my ears when she did. Although we knew it should, we really were surprised when it worked. I was convinced either the curling iron would blow up or we would put London into the biggest blackout since World War II. No problem. Aleen turned beautiful.

5. Got back to the hotel at a decent time but still trying to span the six-hour time gap from Houston to London.

6. Called Ella Part to find out how Antony was getting along. Doing great. It was a heart attack, but a mild one. Both are disappointed they could not meet Aleen. When we return to London, we hope to see them.

7. Hotel not bad. Very close to the airport. Apparently the glide path goes through our room.

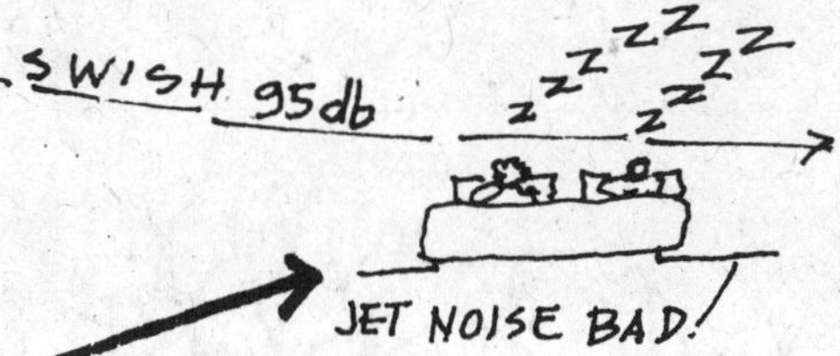

18 January 75 - London

1. This morning Jim Haecker, our fearless leader, and his colleague in the National Bureau of Standards, Bill Cullen, included Aleen and me in their motor trip to St. Albans. (They rented a Fiat.)

2. Rain and fog didn't dampen our enthusiasm for this beautiful old English village with its narrow streets winding up and down. Some were so narrow only one car at a time could make it.

3. First stop was the Black Lion Hotel and Restaurant on Fishpool Street. Unusual name? You should see the street. Some of the centu-

ries-old houses were hanging over it. Some were leaning from old age. Wonderful visual unity, but no monotony.

KINGSBURY WATER MILL

4. Jim and I ordered tea. The waitress brought a big pot crammed with over-sized tea bags, a pot of clear water, and a pitcher of milk. Jim did the honors of pouring the tea -- into my cup first. "No, no, no," cried the lady manager, running over to our table, "one pours the milk first." So I poured my tea back into the pot to do it right: Milk in first -- 1/3 or 1/2 full -- then strong, black tea poured in slowly till the right brown color appears. Then brown sugar. And stir. Delicious!

5. I asked the nice woman if the pot of clear water was to weaken the tea. "No, no, no," she replied, "that's to replenish the tea from the pot full of tea bags." All these 100 years I have used the hot water wrong.

6. Great tea break. Then we walked around the corner to the Kingsbury Water MILL Museum, a working grain mill (they call it a corn mill, but it was supposed to grind all grains but corn). Jim Haecker had been there before and I believe Bill

Cullen had too since St. Albans is where the Building Research Station is located. The BRS and NBS have cooperative programs and both Jim and Bill have spent time at the Building Research Station.

7. Next we went to the open-street market which was open on Saturdays. Despite the bad weather, the roped-off street was packed. Every sensible person carried an umbrella. Not us. Aleen got the "wet look." Glad she brought the curling iron. And was it cold. Brrr.

8. Then on to St. Albans' Abbey. Albans was a Roman pagan converted to Christianity -- executed for saving many people. Some of the materials from Roman walls were used to build the church. Serves them right.

9. How green the grass and hedges were. A few trees even had blossoms. And here it was mid-winter. We were shivering from the cold rain.

10. Passed a shopping area with a large sign reading "Fruiterer." Hurried by that store.

11. Back to St. Albans' Church. There were

THE **WRITTEN HISTORY** OF EGYPT IS ABOUT 5400 YEARS OLD--OLDER THAN THAT OF ANY OTHER COUNTRY; HOWEVER, EGYPT EXISTED LONG BEFORE ITS HISTORY WAS FIRST WRITTEN.

beautiful bronze plates flush with the floor. Had a professor friend at Rice who made rubbings from these. Understand you pay a certain amount an hour for the privilege of making a rubbing.

12. The four of us had lunch at the Black Lion Pub. Were served by a cheery, solicitous English girl. I had trout and the rest had good steak pie. The vegetables served with it all were boiled potatoes and leeks (looked like a cross between onions and celery). Dessert was green apple pie served with what the waitress called double cream, a cream so thick it would hardly pour from the pitcher.

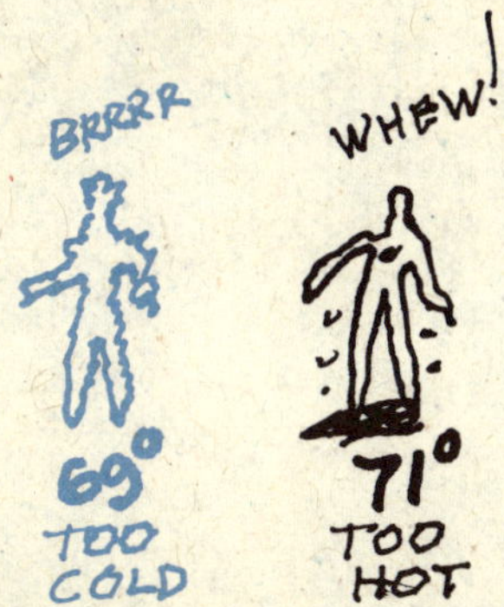

13. Still rain and fog when we left St. Albans for London and the British Museum. Jim drove. Bill navigated. Aleen and I monitored the heating system from the back seat. Jim decided we had a 1° tolerance. He would say, "Now, are you too hot? Or too cold?" I must give him a commercial. He drove in that London traffic like a New York cabbie going on the wrong side of the street. I have nothing good to say about Bill Cullen's proficiency in navigation. We got lost eight times. But we did finally get to the British Museum.

14. No parking place. Our fearless leader stopped the car, got out, went to the gatekeeper, then came back with permission to drive through the gates. Jim literally exuded charm -- a complete change of character. But anything for a parking space.

15. Our purpose for going to the British Museum was to bone up more on Egypt. The Egyptian sculpture gallery was superb. The four of us were amazed at the collection and the effective way of displaying the beautiful pieces. What a civilization Egypt had! What art consciousness! Egypt really has a history.

HEAD AND BODY (OF ANCIENT-EGYPTIAN **FORM OF ART**) ARE DEPICTED FROM THE SIDE, SHOULDER AND EYE.

<u>19 January 75 - London</u>

---

1. Must be nervous. Anticipating the trip. Woke up this AM -- wide awake. Looked at my watch: 6:30. What a short night. Got up, went to the bathroom, turned on the light. Happened to glance at my watch again. Hell! Had it on upside down. It was only 1:30. GAD!

2. Never did plan to tell Aleen, but it slipped

out when I confessed my stupidity. Told her about the time my watch hands became uncoordinated (the minute hand would not line up to 12 when the hour hand was on 3).

3. It bugged me for two weeks until I finally found a good jeweler worthy of fixing my fine electric watch. After he looked at it, he said, "I don't see anything wrong with it. The hands line up OK." Disgustedly I said, "Damn it, for two weeks the minute hand hasn't been at 12 when the hour hand was at 3." "Sir," he said, "you've been wearing that watch upside down for two weeks." I hate myself.

4. Now that we're on the last leg of our journey, I've been thinking more and more about the country we're going to visit. I'm obviously not a student of either ancient Egypt or modern Egypt. But I have always been fascinated with the world of the Arabs, and specifically with Egypt. The other day when Aleen and I were rummaging through some of my things, we ran across a sketch I made 38 years ago, the year I graduated from Oklahoma State University (then Oklahoma A&M). It's an airbrush-and-tempera drawing of the desert and the Nile. For the life of me

I can't recall why I did it or what I copied it from. A photo or a sketch? I suspect it was one of those "a little here, a little there" things. In any case, we decided to have it framed. Looks good and most timely for this trip. Aleen kids me. She says, "Its only value is its antiquity." Remarks like that cause criss-crossed eyes and corrugated mouths, as Willie Pena would say.

WE'RE USING THE SKETCH. **SEE COVER.** SAND, SKY, WATER AND GREAT OLD BUILDINGS— THAT'S **EGYPT.**

A CRS FOUNDER

5. Although neither Aleen nor I has ever been to Egypt, I was at the Cairo airport years ago on the way to Saudi Arabia as a member of the CRS Squatters Design Team for the internationally known College of Petroleum and Minerals in Dhahran. What a great project that has been. There was a large colored photograph of it in this week's Time magazine.

THE ARAB EMPIRE 715 A.D.

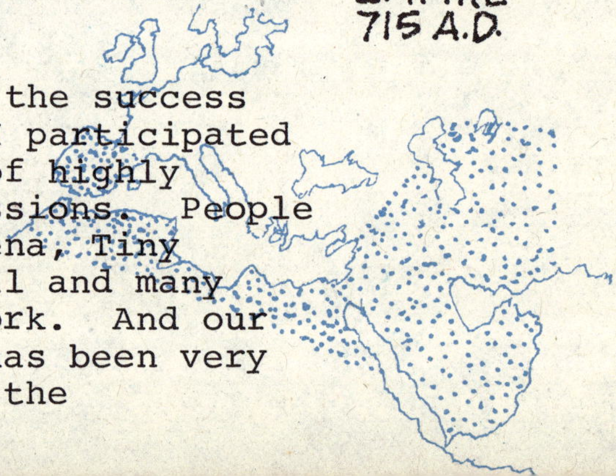

6. I wish I could take credit for the success of CPM, as we call it at CRS, but I participated in only one of literally hundreds of highly concentrated programming/design sessions. People from CRS like Joe Thomas, Willie Pena, Tiny Lawrence, Terry Tengler, John Powell and many others did unusually outstanding work. And our experience working with the Arabs has been very good. I am anxiously anticipating the

ABOUT EGYPT AN INSERT

Been reading a most informative book, Area Handbook for the United Arab Republic (Egypt), Oct., 1970, prepared by the Foreign Area Studies (FAS) at the American University, Washington, D.C. It's full of goodies like these:

Modern Egypt has historical records dating back five millennia (Wallie Scott and you other Aggies, that's 5,000 years). Gives a displaced Okie living in Texas, like me, a complex. My Oklahoma culture is less than 100 years old. And the Oklahoma architectural heritage isn't much to brag about either. But Egypt's architectural heritage. Wow!

Egypt's only legal party is the Arab Socialist Union (ASU). The president is nominated for a six-year term by 43 votes of the National Assembly. Population reported to be 32.5 million in mid-1969, more than 1/2 under age 25. My guess: Today Egypt is about 35 million. Almost all the people are Egyptian Arabs except for Arabized Nubians and

Sudanese (about 50,000 living along the Nile River in southern Egypt near the Sudanese border). About 96 per cent of the people live in the narrow Nile Valley and delta, less than 5 per cent of the country area.

The economy is "mixed," directed by the state since nationalization in 1961; the private sector is limited to small shops (up to five workers), handicraftsmen, and retail traders. Individual income is virtually limited by progressive taxation, forced savings, and compulsory contribution out of wages and salaries for defense purposes. Industry? Shift is away from traditional agricultural processing toward manufacturing. Exploitation of oil, natural gas, and mineral deposits. Oil is the most important mineral resource. Labor? At the end of 1969 the labor force numbered 9.7 million, 29 per cent of the population. Fifty per cent were engaged in agriculture; 11 per cent in industry; 10 per cent plus in trade and finance; 7 per cent plus in personal services (I wonder how many architects Egypt has); 4 1/2 per cent in transportation and communication; and here is the important one coming up, 2.8 per cent in construction. That sounds low to me. I'll check it out when I get there.

About exports: Cotton is a biggie. Petroleum too. Exports go to more than 100 countries. Communist nations (principally the Soviet Union) receive roughly 50 per cent of the total value. Imports? Cereal grains, other foodstuff, machinery, equipment for capital industrialization programs and defense. Communist countries supply about 50 per cent of the total. Money? Basic unit is the Egyptian pound, non-convertible paper currency with multiple exchange rates. There are 100 piasters to the pound. Public debt is rising. Join the club.

professional interchange in Cairo.

7. Tiny Lawrence, who led the design effort on CPM, made a careful study of Arabic customs before he made even one sketch. I asked him to give me the benefit of this Saudi experience so that I might apply it to my upcoming Egyptian experience.

8. Tiny said this: "Bill, as you recall, we started work on the College of Petroleum and Minerals in December, 1964 -- it's been a decade. The spirit and architectural character of the campus is derived from several concepts: Client desire for image from the highway and the use of Saudi Arabian material and Arabic architectural character. We used Saudi sand, aggregate and cement to make an all-Saudi concrete, sand-blasted to expose the natural color. Buildings were arranged along the ridge of the jebel (hill) to create that image and also to define an oasis on top. Columns for all academic buildings are generated from the pointed arch -- the major architectural motif attributed to the Arabic countries. We did, however, incorporate the visual and sonic delight derived from a network of narrow waterways linking major fountains, all

inspired by the rich Moorish environment of the Alhambra."

9. Lawrence adds, "The Arab's way of life seems to be more contemplative than ours and concepts of beauty and spirit seem to be a natural part of their make-up. Although the Arab client is interested in efficiency and analysis, there is never a need to disguise a discussion on concepts of architectural form and beauty."

10. Buildings are never fully successful unless the users love their building. In those terms, CRS' first desert building for desert people is successful. Last month, Tiny received a letter from Dr. Bakr Abdullah Bakr, Rector (the President in our terms) of the College of Petroleum and Minerals, substantiating the users' delight -- the true measure of success.

11. Dr. Bakr writes, "Dear Mr. Lawrence, a few hours ago His Majesty, King Faisal, dedicated the new campus of the College of Petroleum and Minerals here in Dhahran. Ministers of state, representatives of the governments in the Arabian Gulf area, the diplomatic corps, distinguished educators, trustees, our faculty and students,

THE **ARAB** APPRECIATES THOSE WHO RESPECT HIS TRADITIONS AND HIS MANNERS. HE BEFRIENDS EASILY THOSE WHO UNDERSTAND HIS WAY OF LIFE.
ARABIC SCHOLAR

and representatives of the giant industrial enterprises of this kingdom were present. It was a day of fulfillment of our hopes and dedication to the solemn duties of the future. It is dedicated not only to serve the Government and the industries of Saudi Arabia, but the youth of this kingdom as well." (On my return, I learned King Faisal has made the College a University.)

12. Dr. Bakr continues, "As we walked through those beautiful arcades, sat in some of the little theaters, or visited the classrooms and laboratories, our thoughts went back over the past 10 years to the time when this was only a dream, a concept in the minds of a few dedicated persons. This new campus will be identified with the economic renaissance of this kingdom. We thank all of you for your contribution to this goal." Beautiful. He writes like Hassan Fathy.

13. Projects like this make me glad I'm an architect. And particularly glad I'm on the CRS team. Without a team, I would be nothing.

19 January 75 - London - At Heathrow Airport

1. London's Heathrow Airport really is buzzing, particularly with international travelers. I understand that there are two good reasons for this: a) the pound is good for monetary exchange in most countries, and b) all airways lead to London. The Skyway Hotel was full of foreigners, particularly black Africans.

2. Got a big kick yesterday at the British Museum. Aleen and I were looking at some unwrapped mummies. Two little black boys were talking excitedly about the human remains. One boy said, "Awful! They scare me." The other said, "You know they're not real; they're here just to fool us." They were real -- and over 4,000 years old.

3. The Egyptian sculpture was magnificent. Some pieces were superb using any evaluative criteria, any time in history and in any culture. They came from an advanced civilization over 3,500 years before Christ. Has man made progress? Artistically, certainly not much.

4. I noticed the later-day pieces were over-

FIONA PATTERSON OF CRS DESIGN ASSOCIATES ONCE WORKED AT THE **BRITISH MUSEUM.** SHE SAID, "WE CALLED ONE OF THE MUMMIFIED CORPSES **'GINGER JIM'** BECAUSE OF THE REMNANTS OF RED HAIR ON HIS HEAD. **GINGER JIM** CAUSED MORE COMMENTS THAN ALL THE OTHER EXHIBITS OF THE DEPARTMENT OF EGYPTIAN ANTIQUITIES PUT TOGETHER."

embellished and, in my eyes, not nearly as good as the earlier things. Someone once said, and I believe it, that the first sign of decadence is when a country's art (and lifestyle, too) becomes over-embellished.

5. Waiting room (debarking) at Heathrow not really bad. Comfortable chairs. Great duty-free shops. Adequate graphics and announcements to let you know where you stand on time/place of your departure. And nice, clean toilets.

6. Aleen was rather shocked at an incident in the little girls' room. To put it in her words: "The English lady attendant had to be very firm with a Pakistani woman with her baby. The attendant repeated several times before the woman understood, 'You don't wash the baby's bottom in the sink.'" Not in my sink, she wouldn't.

7. And speaking of bottoms, this morning yours truly was stricken slightly. Aleen had the Lomotil, but would she give me one? No! She said, "Let's wait 30 minutes." On this trip those pills have greater value than dollars and oil.

Aleen wasn't letting me spend anything foolishly. It's not oil that will control the world; it's Lomotil.

19 January - Enroute to Cairo

1. A few minutes ago we took off for Paris on Egypt Air. Believe we're in a 707. Not quite sure. Anything bigger than a Cessna confuses me.

2. Jim Haecker and Bill Cullen blocked off six seats at the emergency exits for the four of us -- allowed us better than first class leg room. Very comfortable.

3. English towns from the air different from U.S. Patterns reflected row housing, closely clustered. No urban sprawl. And no FHA planning forms, thank goodness.

4. Just flew over the shores of the English Channel. Great cliffs. And white. Are these the white cliffs of Dover?

5. Didn't take but a few minutes to cross the

Channel. Quite a few cliffs on the France side too.

6. Landed at Le Bourget Airport, Paris. What happened to Orly? Never landed here before. Remember now; it's where the international air show is held. Jim tells me there are three airports in Paris. Bill Cullen is still with us. What a nice guy he is.

7. Forgot to mention when we were going through security at London's Heathrow, I was literally frisked. I was wearing the new cashmere coat which Aleen recently gave me. When the man frisked me, he remarked, "That coat sure is of quality -- so soft." I had no comment except to blush. Aleen heard his remark and added, "So that's why they've been overattentive. I thought it was you. It's your coat." At my advanced age, I'll take any points I can get. A woman frisked Aleen. It had better have been a woman.

19 January - Cairo

1. We're finally here! At the airport we got the VIP treatment. A young man from the Ministry of Housing and Reconstruction, a wheeler-dealer from way back, took us through customs, health and money exchange. With minimum stress on our part. He will continue to serve as our pusher. He has a driver and a minibus. Strictly first class. His name is Nabil Ahmed.

Another greeter was an old friend of Jim Haecker's, a beautiful person who sparkled with wit, Hammed Kaddah. Also there to help us was Dr. Ahmed El-Erian. Dr. El-Erian is the coordinator of our joint study group, Haecker's Egyptian counterpart.

2. We rode the minibus into the city. Lots of traffic considering it was after midnight. Cars use only dimmers.

3. About a 40-minute ride to the Hotel Nile (last-minute change).

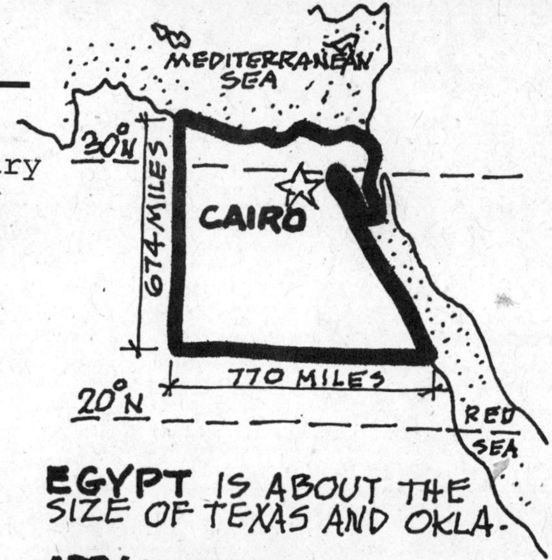

AREA: 386,000 SQ. MI.
POPULATION: 35,000,000
CLIMATE: HOT SUMMERS, MILD WINTERS.
ELEVATIONS: 7,000 FT. TO 100 FT AT CAIRO☆. SOME AREAS INCLUDING THE QATTARA DEPRESSION ARE ALMOST 450 FT. BELOW SEA LEVEL.

20 January 75 - Cairo

1. I was sick throughout the night. Even with the miracle international pill. Aleen issued me one at 3:30 a.m. with the perceptive words, "She who controls the Lomotil controls the world." I had a miserable night. No sleep. No privacy. No drinking water. And a bad case of London's "low blow." I think it came about because of a big piece of green apple pie and double cream. A plain case of overindulgence.

2. We had a terrible room. Had to share a bath with another couple. Didn't find out who they were until this morning after we checked out. It was a fellow team member, Phil Bush, and his wife, Margie.

3. Despite the bad room, when Aleen and I opened the wooden blinds this morning we were given a beautiful view of the Nile. Great! Beautiful river. The cityscape carried a variety of building forms. No monotony.

4. It's going to be a heavy day. Hope I start feeling better.

IN OUR BLISSFUL IGNORANCE WE ARE SPELLING IT THIS WAY-- **SHEPEARD HOTEL.** IF YOU WANT TO SPELL IT LIKE IT'S ON THE POST CARDS--SHEPHEARDS HOTEL--THAT'S YOUR BUSINESS. WE'VE SEEN IT SPELLED 6 DIFFERENT WAYS.

5. First thing we did was to change hotels. Moved to the Shepeard's. Much, much better. Also a great view of the Nile. And a big private bath with two sinks. Now Aleen can't fuss at my encroachment.

SHEPEARD HOTEL
ON THE NILE AT CAIRO

6. Our group was minibused to the Academy of Scientific Research and Technology. Dr. Abdel Moneim Abou-El-Azm gave us a cordial welcome and a thorough review of the building materials and technology situation in Egypt. He's very sharp, but went a little fast for me. I asked a few dumb questions like: "What are we supposed to be doing?" Nevertheless, in these kinds of meetings, the question is basic. And I really didn't know.

7. Then we met with the Egyptian counterparts of our task force. The updated listing is:

Prof. Dr. Ibrahim A. Gabar, Head of Structural Engineering Dept., Faculty of Engineering, Cairo University;

Prof. Dr. Ahmed A. El-Erian (Coordinator), Professor of Materials, Civil Engineering Dept., Faculty of Engineering, Cairo University;

THE LIST WE WERE GIVEN FAILED TO INCLUDE ARCHITECT **HUSSEIN SALEH** WHO LATER CONTRIBUTED GREATLY TO OUR EFFORTS.

Engineer Ismail I. Osman, Technical Director, Minister of Housing and Reconstruction Office;

Prof. Dr. Hassan K. Imbabi, Head of Building Materials Dept., National Research Center;

Dr. Eng. Mohamed A. El-Masry, Undersecretary, Ministry of Housing and Reconstruction;

Dr. Mohamed H. Ramez, Head of Raw Materials and Processing of Building Materials Dept., Building Research Center, Ministry of Housing and Reconstruction;

Prof. Yousef Shafic, Head of Architecture Dept., Faculty of Engineering, Cairo University.

Also with us were:

Dr. Kaissar M. Hanna, Assistant Professor of Research, Building Materials Dept., National Research Center; and

Dr. and Mrs. Doreyo M. Shohim, Assistant Professor, Building and Refract. Materials Dept., National Research Center.

8. Some rather strong opinions were put on the table by our Egyptian friends, experts in their own right. Dr. Gafar pointed out that the main problem is how best to use local raw materials, of which there seem to be plenty.

9. Coordinator Dr. Ahmed El-Erian came out in no uncertain terms saying that Egypt should stop robbing the topsoil to make bricks.

10. The problem is simple. The solution very complex. For 5,000 years Egyptians had been making bricks (from the silt of the Nile) a certain way and using them in their buildings. Just about all buildings. Now along comes the Aswan High Dam, 600 miles away, to create badly needed electricity and water for irrigation. But in doing these good things, the dam stopped the silt from flowing down the Nile. No silt. No bricks. A real problem.

11. Having read Fathy, who advocated mud brick, I asked if it's used much. "Very little," Dr. Hanna said. "Ninety-nine per cent of all brick is fired."

DOES FATHY KNOW THIS?

12. Dr. Ramez was hot on the use of expanded

sand and lime brick. Said it was as strong as brick from the Nile.

13. Someone asked about the feasibility of concrete blocks. Too expensive at this time, they told us.

14. And what about precast, prestressed concrete? Problems of equipment and transportation. This was vividly pointed out when, on our way minibusing from one building to another, we saw a donkey-drawn cart carrying angle irons.

15. My stomach isn't the only thing about me that's confused. So is my brain. We visited many bureaucratic organizations with similar sounding names like Building Research Institute, National Research Center, National Academy of Scientific Research and Technology. Egyptians, like Americans, change their names to inaugurate slightly different programs to give the impression of a fresh start. Always did get hacked off when educators tried to change the name of the Library to Educational Resource Center to do away with the connotation of a library as a book morgue. Vestiges still remain. In any case, I'm confused by changes of names which have occurred.

16. Jim Haecker has been doing a masterful job of diplomacy, logistics and "sorting out the issues," as he so aptly put it. There is so much confusion -- overlapping authorities/responsibilities of the different governmental agencies. Even jealousies. And there is that same old "gown and town" fight going on between Professors in Academia and Practitioners in the Real World, only more so here in Egypt.

17. Certainly there are more professors per capita in Egypt than in the U.S. because everybody gets a free education. Cairo University has 84,000 students. We were told that there are 4,000 Ph.D's in the agricultural department alone. I believe Egypt has eight universities. Six offer degrees in architecture. At the reception tonight I met one architect, Dr. Hussein A. Saleh, who went on to get his Master's degree from the University of California at Berkeley and his Ph.D from the University of Pennsylvania.

18. Our meetings during the day were most friendly and open. Plenty of dialogue. Hopefully they will prove productive. Our agenda was considerably over-structured, leaving no time for thinking. I'm over-saturated.

The devil (Bill) made me do this.

19. Before the official reception this evening, Aleen went to the hairdresser's across the street from the hotel. Says she, "I was under the dryer when I saw this handsome young man (23 or 24) come in from the adjoining barber shop. He was dressed in very mod fashion -- black flared pants, yellow shirt and a black-and-white pull-over sweater. Two girls were blow-drying his hair while the gentleman hairdresser was styling it with a brush. I thought he must be a sheik or a movie star. Turned out to be just a pilot for the President of Libya who was visiting Cairo."

20. I never got that much attention. And I fly, too. Also have long hair. However, I did get a buzz when one of our hosts referred to Bill Cullen (with his dark head of hair) as Black Bill and me, Bill Caudill, (with my white mane) as White Bill.

21. Dr. Abou-El-Azm honored the American delegation at the reception. Very nice. And he was most gracious. My problem, however, was the inability to say "no thank you" to all the cookies, cake and juices. Who am I to create an international incident by refusing hospitality?

The Wee Hours of 21 January 75:

1. Lomotil ain't the miracle some people claim.

2. I'm back to reading Fathy's *Architecture for the Poor*. His argument for mud brick walls continues to be strong; however, he does concede that "thick mud brick walls are not a perfect means of keeping cool, for although mud is a poor heat conductor, it retains heat for a long time."

3. Fathy insists that heat-resisting qualities of mud brick walls are better than concrete and calls the reader's attention to the concrete houses at Kom Ombo, built by the sugar company for its employees. The houses were too hot in the summer and too cold in the winter. The employees preferred to live in the mud houses of the peasants.

4. Since mud brick walls retain the heat of the day, it's much hotter at night inside than outside. According to Fathy, the obvious solution is to live downstairs during the day, then at night move onto the roof to sleep in the cool night air.

5. Fathy likes the courtyard. It serves as a well into which the cooler air from the roof sinks, cooling the downstairs during the night.

6. Fathy strongly advocates natural ventilation of houses, too. I wonder about this. I wonder if the air isn't too dry. Our experience in West Texas proved that evaporative cooling (introducing moisture into the house) was much more effective than natural ventilation. Hot winds are unwanted. But moist breezes from the Mediterranean, I'm sure, must be welcome.

7. Fathy discovered by accident what we proved in the wind tunnel at Texas A&M University in the early 50's: that the larger opening should be on the leeward side and the smaller openings on the windward side.

8. Goodnight, Fathy.

21 January 75 - Cairo

1. For me this is "take it easy" day. Three days of Lomotiling have taken their toll. Great way to lose weight, but no fun. Oh! to eat again in Romana's Cafeteria in Houston.

2. Still on my problem: I suppose I should confess my remarks at a meeting at the University of Cairo with the President and his group. One professor made a serious plea for U.S. support to Egypt -- to the effect that we both believe in God and we both know that God will punish us (the U.S.) if we mistreat smaller nations like Egypt. Looking me straight in the eye, he ended rather passionately with, "And he will punish you." I retorted, "He already has." Everyone in the group knew I was taking Lomotil.

3. One more irrelevant bit of trivia: It's been rather chilly here most of the time, but not bad. Cairo in winter is quite a spot.

4. Today is also architects' day. Up to now, most of the people we've been in contact with have been a variety of engineers and scientists. Today I met Yousef Shafic, Professor of Archi-

tectural Engineering -- a very nice guy. Did graduate work at U. of Illinois. Worked for Pace Associates, Chicago. As I understand, my friend Architect Dr. Hussein A. Saleh is also a top administrator in the Department of Architecture, Faculty of Engineering, Cairo University.

INFORMATION GIVEN TO US BY THE CLASS A'S.

5. There are three kinds of architects in Egypt: Class A (licensed), about 5,000, about 400 per year coming out of the schools; Class B, fine arts types; and Class C, other kinds. Classes B and C are the draftsmen.

6. Where do they work? Fifty per cent find better deals in other Arab countries.

7. All young men are drafted into service for two years. Some have been in five years. No retooling program.

8. Actually the building industry is in a mess. It has dropped tremendously over the last 20 years. And Egypt has been at war for 20 years. Fifteen years ago there was a nationalization of the building industry. It's not working. Now the government is being more open and encouraging to private enterprise.

9. But there's no quality control or guaranteed performance of building materials and systems. Last month an eight-story apartment building (less than a year old) collapsed. Bad structural design and poor quality of materials.

10. Jim Haecker remarked, "The process supports poor quality."

11. My hero, Hassan Fathy, is not Egypt's hero. They say Fathy didn't help many poor. He only did two projects. "Gourna Village," they said, "is a spook village." People didn't like to live there. "Fathy imposes how he thinks these people should live, not how they want to live." A prophet is without honor in his own country. However, I understand he was used as a consultant for projects in Saudi Arabia.

FATHY DOESN'T WIN THE POPULARITY CONTEST WITH SOME EGYPTIAN ARCHITECTS.

12. Building problem in a nutshell:

a. Don't have labor (skilled and semi-skilled)
b. Don't have materials

All windows are wood, imported and poor quality, tailor-made. Glass hard to get. Plumbing and electrical imported. There is a lot of gypsum,

PREPARED BY FOREIGN AREA STUDIES OF THE AMERICAN UNIVERSITY

but no dry-wall construction.

13. According to the Area Handbook for the United Arab Republic, there is raw material for cement and concrete construction. Then why are there so few cement plants? We found no definite answers. Jim and I feel that a strong contender for help to solve the "wall problem" is the use of concrete block. Easy to lay. Easy to produce, with reasonably priced equipment. Sounds too simple.

14. One interesting bit of information from the Handbook: Despite the fact that the country was in need of skilled construction workers, the government permitted 5,000 laborers to work in West Germany in 1964 and in subsequent years. According to official views, the main motives in granting this permission were the need for foreign currency and the continued training of the participating workers. It seems like a logical plan to me. Rather like the exchange student programs we have in this country.

15. This is for sure. The Egyptians we have talked with are convinced that a live, meaningful construction program is the hope for Egypt.

16. One professor said attitudes must be changed. "We've been in a hurry for 20 years. Quality has suffered."

17. I continue to ask myself why we are here. How can we best help these people? I'm sure that all some of them want is money, not advice. Others want us to sell their superiors on the idea that what they are doing should get stronger official support.

18. Each prof feels that research is needed, but research is wasted without pilot programs.

19. I talked earlier about Dr. Abou-El-Azm, who gave us a fine welcoming speech and an outline of the building industry in Egypt. I had heard of "worry beads" before, but never saw them in action. Dr. Abou-El-Azm fiddled with them during his talk. I wasn't quite sure exactly what they were, so naturally I asked. He explained the religious meaning and why they were called "worry beads." I asked, "When you count them, do you say 'Damn Americans, damn Americans, damn Americans'?" Sounds like a great therapeutic device. And who gave me a set? None other than my newly-found architect friend, Dr. Hussein A. Saleh.

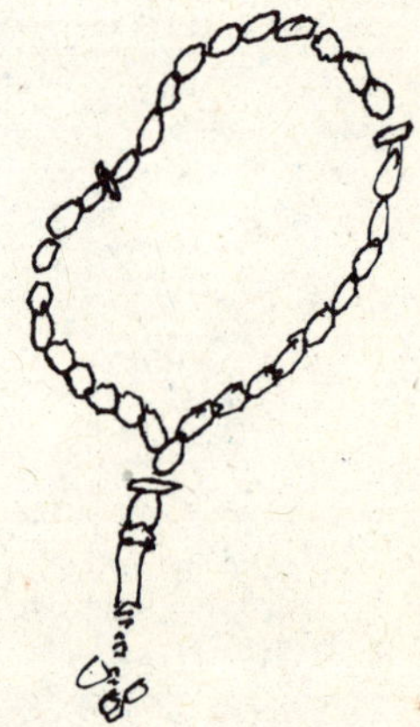

Architects all over the world seem to like each other. A true minority group. Must stick together.

20. Egypt has more than war troubles (threats of invasion) and brick shortage problems. There's a "cotton worm" which costs millions of pounds a year. AND A fruit fly that's playing Old Ned with the fruit crop.

21. I fasted today, trying to break this damn unpleasant new habit. Ate rice at noon. Best taste since I ate that green apple pie with "double cream" in St. Albans, England (I know that did it). Tonight I feel a bit better. Aleen and I walked over to the Hilton and ate a snack at an Americanized coffee shop. It had a microwave oven taste but it <u>was</u> American. All the food here tastes and smells the same.

22. The Sheepeard's Hotel is more Egyptian than the Nile Hilton, and there must be 10 times more servants all smiling genuine smiles. (I wonder if they really are.) I smile back. But I'm afraid that won't hack it. Money smiles best.

23. Going to bed feeling a lot better. My

by Anna Ragon

C. It produces only high tides, and no low ti

D. It produces only low tides, and no high tic

problem is subsiding.

22 January 75 - Cairo

1. It's a great day. Lomotil is living up to its name of miracle drug. But it sure took its sweet time.

2. There's a little man on our floor who seems to serve as the housekeeper. He smiles so much he scares me. He likes me. I trade him American quarters for Egyptian money. Give him a good deal. When Aleen and I get off the elevator, there's that smile. When we open the door to leave our room, there's that smile. It's one big race to see whether I can beat him in opening my door or punching the elevator button. He's a tough competitor, smiling, smiling, all the time. This morning before we left the room, I cautiously stuck my head out the door and then quietly called back to Aleen, "All clear, honey; the little man isn't in sight." With confidence we strode down the hall toward the elevator. But there he was, lurking in a side hall, waiting for us with his three-word English vocabu-

lary, "Good morning, please." The smile strikes again.

3. So did the Lomotil, damn it. Aleen is still okay. Women are the weaker sex? Ha! Ain't true.

4. Egypt's weather is predictable. The political climate is something else. And from my point of view, it seems to change all the time. If not day to day, then week to week.

5. In this morning's paper, The Egyptian Gazette, a very disturbing editorial appeared entitled "The Right to Murder." Let me give you a few excerpts: "If, therefore, as is expected, the U.S. pours out more money to arm Israel, it can only mean that Washington does not envisage peace and is in fact encouraging Israel to use its arms in offense to thwart peace.

"The U.S. thus arrogates to itself the divine and unquestioned right to arm Israel to the teeth. It also arrogates to itself the right to allow Israel to use U.S. weapons to kill Arabs and destroy their land. Despite this, the U.S. arrogates to itself, in addition, the divine and unquestioned right to demand Arab oil at lower prices and to have Arab money recycled to bolster western economy. Equally serious is the fact that the U.S. arrogates to itself the right to threaten the Arabs with force and occupy their oil wells.

"This is the U.S. policy as White House actions give us to understand. If this is the case, how, in the name of sanity and sense, can U.S. policymakers and those who manipulate them then expect the Arabs to continue to have faith in them and their good intentions."

You wonder if our Egyptian professional colleagues are trusting us.

6. Back in Houston on 14 January, our morning paper made it clear that there was an opportunity to negotiate a Middle East peace settlement. Only the week before, things had looked very black. The new hope was attributed to Russian leader Leonid Brezhnev's decision to postpone his scheduled visit to Egypt -- which was to occur concurrently with my arrival, by the way. I leave myself open to those wits on the CRS team who will say it's a good thing --

Egypt couldn't stand two demagogues at the same time.

7. January 13's <u>Christian Science Monitor</u> (I love that newspaper) ran an interesting story on Mr. Brezhnev and a lotus flower. According to the article, post-dated Aswan, Egypt, the postponement of Brezhnev's visit to Egypt took "a bit of the bloom off the 235-foot-high concrete-and-marble lotus flower on the edge of the Soviet-built Aswan High Dam." The giant lotus flower, symbol of ancient Egypt's Upper Kingdom, is meant to be Egypt's way of thanking the Soviet Union for its 10 years of technical, manpower and financial help in building the billion-dollar dam, Egypt's largest source of electric power. Mr. Brezhnev, the Soviet Communist Party leader, was scheduled to dedicate the monument on 15 January, the fourth anniversary of the dam's completion.

8. During the last few months before Brezhnev's contemplated visit, workers were put on double shifts to finish the huge monument. Designed by two Russian artists, it has been in the works for more than three years, at a cost of $3.7 million.

According to the article, the lotus now sits amid a bleak, rubble-strewn bed of sand and gravel on the western edge of the Aswan High Dam. Someone was disappointed, I'm sure.

9. The Houston Post article (14 January) which talked about a peace settlement also reported that the Shah of Iran had offered to make up the annual six million tons of oil which Israel would lose if it handed back the Abu Rudeis fields to Egypt. Iran already supplies Israel with nearly 50 per cent of its oil needs. Israel is getting the remainder from Sinai.

10. Apparently the position of Syria remains crucial. Whether Egyptian President Sadat can accept a new bilateral agreement with Israel without some corresponding agreement on the Syrian front is problematical.

11. Iran holds a key position in the Mid-East peace role. It has growing oil and military power. The latest report is that Shah Muhammad Reza Pahlevi may be seeking to stimulate new Arab-Israeli peace efforts, according to Egyptian diplomatic sources.

EGYPT IS IN THE MIDDLE.

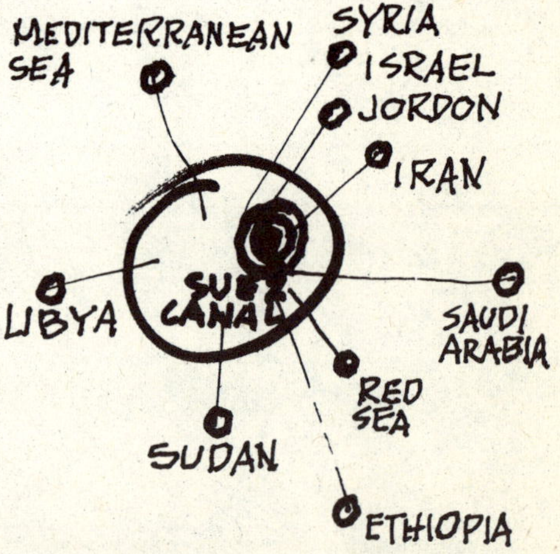

STRATEGICALLY LOCATED, THROUGHOUT HISTORY EGYPT HAS BEEN THE CONTROL POINT FOR BOTH COMMERCE AND CONFLICT.

12. In January, the Shah of Iran and his Empress Farah visited Cairo. Reportedly President Sadat praised Iranian support for the Arabs at that time. He referred to a statement issued after the Shah's visit to Amman, rejecting Israeli changes in the status of East Jerusalem and advocating recovery of occupied Arab land and the "legitimate rights of the Palestinian people."

13. During the visit to Egypt, the Shah gave a spectacular gift to Egypt -- a fleet of 1,000 brand-new Mercedes buses for Cairo's overloaded city transport system, one of the causes of riots in the city on New Year's Day.

14. I'm ready to be picked up to see buildings. Not sure my poor old stomach is.

22 January 75 - Cairo - Before Lunch

1. Architect Hussein Saleh arrived promptly at 9 a.m. Drove us to Architect Yousef Shafic's apartment so he could join us. He lives in a 40-year-old brick stucco house, nicely designed;

original workmanship superb. I was told workmanship with that high quality is impossible to get today.

2. We drove to an 18-story apartment house under construction. Takes three to four years to build. Talked with the constructor and owner. Greatest needs are concrete mixers (trucks preferred), steel scaffolding and elevator machinery.

3. In the high-rise apartment housing, which is about halfway finished, I saw the following: a) some steel forms for slabs, b) mostly wood forms, c) asphalt waterproofing being melted by open flame, d) workers feeding sand and gravel, mixed for concrete on the site, into baskets, e) yellow face brick, f) common red brick from Nile silt, and g) expanded sand/lime blocks.

4. The expanded sand/lime blocks, which some of the researchers are so hot about, are used as fillers with the common brick. Bricks were used at openings and corners and sometimes as ties between corners -- very similar to the way concrete block was used in our country when it first came out.

5. Was told by Architects Saleh and Shafic that the expanded sand/lime blocks (about 6" x 8" x 24")in place cost 1/3 less and weigh 1/3 less than the ordinary commons. One block takes the place of six bricks.

6. I did see some concrete blocks used as fillers in slabs. Poor quality. Poor strength. So Egypt does manufacture some concrete blocks.

7. There is one factory making the expanded sand/lime blocks. Must be doing a booming business since I saw them on many construction jobs (used only as filler between bricks or spandrels). I understand the factory has been in operation less than two years. It's of Polish derivation.

8. But expanded sand/lime block has one great weakness: its characteristic of soaking up water. For example, unless the block is soaked thoroughly, the mortar loses its water and turns to powder.

9. My architect friends (both very nice people and good professionals) told us that before nationalization of the building industry, things

were in very good shape. For example, private industry manufactured gypsum board for dry wall construction.

10. Chip board (softpress fiber board) was manufactured too. Some still is, but very limited in use. Must be too expensive.

11. In the early experimental stages, they built panel-type houses of chip board made from sugar cane stalks. Good material. Withstood the elements. But they left one thing undone. They forgot to take the sweet taste out. The rats ate up the wall!

12. Both Hussein and Joe (that's Professor Shafic -- when he was taking graduate work at U. of Illinois and working in Chicago, they called him Joe) have consulted with and are doing research with the Arab Research and Administration Center (ARAC). ARAC is Middle East and does everything, from feasibility studies for housing developments to locating pickle factories. It's booming since foreign investors and developers are seeking fortunes in the Middle East. ARAC plans to have an office in New York.

13. Both architects were greatly disturbed that Cairo expansion was gobbling up precious top soil for food -- trading houses for food in Egypt where growing land is shrinking is bad business. I wonder what the hell those 4,000 Ph.D's in Agriculture have been doing. Anyone can grow things in a fertile valley. The desert must also be used.

14. We visited a Hussein-designed 12-story apartment building. Walked almost all the way up. Great view. Could see the pyramids. When we looked down the 10 or 11 stories, we saw a camel. What contrast -- an urbanized high-rise and a caravanized humpback.

15. Hussein's 12-story apartments were most unique. Very luxurious in space. Each was two floors with three split levels -- like criss-crossing stairs.

16. We also saw a four-story house designed by Architect Shafic. Beautifully proportioned. Great strength in its simplicity. Mies Van Der Rohe's influence was there. He's a damn good architect -- Yousef Shafic.

17. Both Hussein and Joe complained about poor quality control during construction and waste of materials and labor. "We build the buildings three times," Joe said, implying that many corrections are required.

22 January 75 - Cairo - After Lunch - Aleen

1. Bill asked me to take over for a while. So here goes.

Here I am again. How I ever got into writing memos I'll never know.

2. Our tour this morning began with a maneuver around a huge banyan tree (at least 8' in diameter), literally in front of Professor Shafic's apartment. Cars had to squeeze by one another on the six-foot-wide path on one side of the tree.

3. Hussein Saleh wanted us to have some of the Egyptian navel oranges so we stopped at an open fruit and vegetable market. He selected some oranges for us -- they're very good -- and some very tiny bananas. A native woman was leaving the stand with a huge cabbage, bigger than a basketball, balanced on her head. Bill said his

mother would have loved seeing that. She was crazy about hats and this woman looked as though she were wearing a very fancy one.

4. We were invited for coffee at Hussein's apartment at the end of our tour. He said his wife wasn't home but he knew it would be a treat for me to have some Maxwell House coffee perked in a Cory coffee pot. He had brought the coffee pot from the States and he made the coffee very weak by his standards. Egyptian coffee is strong and black. I can't drink it without milk.

5. We walked up four floors to Hussein's apartment. I was surprised to see it furnished with a shag area rug in the living room and some leather and chrome chairs (very contemporary). Hussein was influenced by his study in the U.S.

FROM HUSSEIN'S BALCONY WE COULD SEE THE 3 GREAT **PYRAMIDS** AT GIZA ON THE EDGE OF CAIRO.

ALL WERE CONSTRUCTED DURING **DYNASTY IV**. ON THE LEFT IS **MYCERINUS** (C. 2500 B.C.). CENTER IS **CHEPHREN** (C. 2530 B.C.). AND RIGHT IS THE BIGGEST ONE-- **CHEOPS** (C. 2570 BC.)

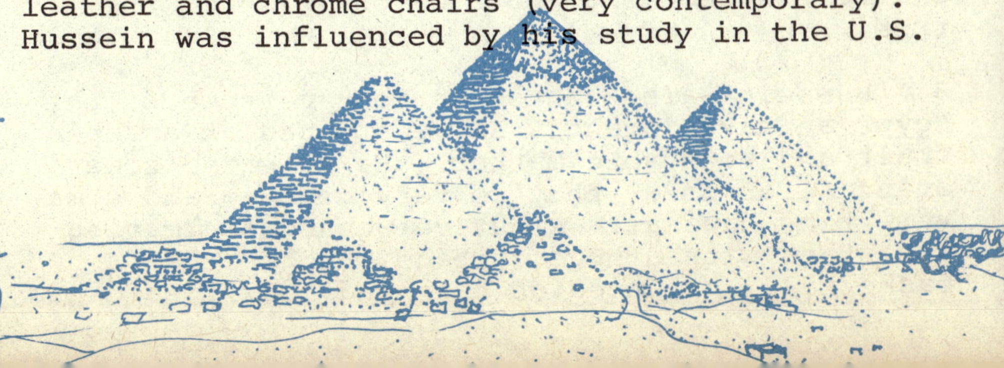

22 January 75 - Cairo - Evening

1. Professor Shafic pointed out that the most uncomfortable months in Cairo are April, May and June when hot, dusty winds blow all the time. Each day, someone figured, eight tons of dust settle on Cairo. This is only January but at least a ton of dust must have settled on me today. Since Aleen didn't get to finish her account of our day's adventures, I'm passing the pen to her while I bathe.

ALEEN CONTINUES WITH THE BLURB.

2. After lunch Jim Haecker, Bill and I joined Jim's friends -- Architect Nabil Gali and his wife, Azza Fahmy, a jewelry designer/silversmith, and her sister, Randa Fahmy.

3. When we arrived at the Galis' apartment for tea, we were greeted by their boxer, Sheila, who had given birth to 12 puppies just a week before. Five of the puppies had died, but Sheila and the surviving seven had taken over Nabil and Azza's bed. For the time being, the Galis are sleeping on the studio couch in the small living room.

4. The Galis' apartment is what I would call a typical architect/artist's living quarters. One room is her shop, another is for his drafting table, and the dining area is used to display her jewelry. The coffee table top was thin copper which had been used as "plates" over the rollers of newspaper presses. Art displays were very good.

5. On their porch, the Galis had some hanging flower pots with beautiful but unusual shapes. They explained that the pots were actually porous clay water-cooling vessels. There is leakage through the clay caused by absorption which lowers the temperature of the water 20°. The vessels made attractive containers for the flowers.

6. Nabil, his wife, Azza, and her sister, Randa, are most gracious, beautiful people.

7. Here comes Bill. I'm relinquishing the pen.

22 January 75 - Cairo - Late Evening

1. I'm back. Clean and feeling good. No Lomotil in 12 hours. Hope I lost about seven pounds from this ordeal.

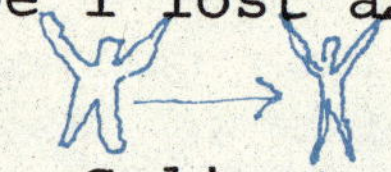

2. This afternoon the Galis took Jim, Aleen and me to a wonderful little village designed primarily for artisans, specifically weavers and potters.

3. While there we met a handsome Egyptian couple from Alexandria: Magdy Farahat and Alfred Lian. He's a nuclear scientist. They had spent four years studying at Northwestern. They liked Chicago and were so pleased to meet us. Invited us to visit them in Alexandria. Also wanted to introduce me to Hassan Fathy, "Egypt's most creative, influential architect." They thought Fathy was the greatest of all. Both of these nice Egyptians were so warm and friendly, and they became acquainted with Nabil, Azza and Randa who are even more warm and friendly.

HERE'S THAT MAN AGAIN!

4. Rather interesting that Nabil Gali, an architect, said he felt Hassan Fathy was too pub-

NEW GOURNA VS. PRUETT IGOE

licity-minded. Did things really more to help himself than the poor. May be too harsh a criticism. Of course not many architects liked Frank Lloyd Wright. Gali's reasoning is that Fathy's pride and joy project at New Gourna failed for two reasons: a) Fathy tried to make the mountain people (historical artifacts hunters) live in the valley, but they moved back to the mountains, and b) the buildings all cracked. The photos I saw were beautiful forms and spaces, but unhappy people can tear up buildings and move out of them, as evidenced by the Pruett-Igoe housing project in St. Louis.

5. The artists' village we visited today was somewhat similar in concept to Hassan Fathy's village. Both were made of mud. Both designed by fine architects. Both for poor people. But Nabil insists that the village we saw today -- Harania Village, near the pyramids -- is designed on solid sociological bases primarily to help people (the users), not to be a monument to the architect.

6. The architect who started this "experience in living," as Jim Haecker so aptly put it, was Professor Ramses Wissa Wassef. Nabil swears by

him and was Wassef's student for four years. (Prof. Wassef died last year.)

7. The story, as I have it, began about 20 years ago when Prof. Wassef went to a small group of peasants to see if there were ways to improve their living conditions. Somehow they got together in making art objects and selling them. As time went on, talents were developed and products became more artistically pleasing. Children were schooled in the crafts and the people began to make some money. Today the products, specifically wall hangings, tapestries and batik, are simply great.

SEEING THIS **VILLAGE OF ARTISTS** WAS ONE OF THE HIGH POINTS OF OUR TRIP TO EGYPT. ALEEN AND I LOVED THE PEOPLE, THEIR WORKS, AND THEIR BEAUTIFUL LITTLE VILLAGE.

8. A new exhibit will open next Friday near our hotel. Aleen and I hope to see it.

9. This month the Smithsonian Institute will have an exhibition from the village. It's called "Egyptian tapestries from the workshop of Wissa Wassef in Harranta, Egypt."

10. If pure results decide the credibility of an architect's work, this village of artists, started by Architect Wassef, should place him among the greats.

THE ENGLISH SPELLING OF EGYPTIAN NAMES HAS REALLY THROWN ME. IS IT **RAMSES**? OR IS IT RAMESES? OR IS IT RAMESSES? **GIZA** OR GIZEH? NOW I SEE THAT THE ARTIST VILLAGE HAS AT LEAST 2 SPELLINGS. YOU DECIDE. I'M TIRED.

11. The villagers build their own buildings including schools, shops, museum, barns and homes.

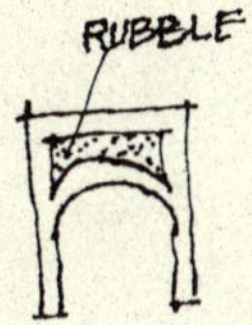

12. The homes are being built in stages -- first floor, first stage. They have stone walls with brick domes for ceilings. Parapets are extended level with the top of the dome, then the space is filled in with rubble -- ready for the second floor.

13. To construct the dome, light steel cage-like forms were used to hold the brick domes in place until they were set. Easily removed. Great insulation. Since the temporary roof was flat, the space between the walls and the dome had to be filled in with rubble. Mass is good insulation. Then, of course, there were the two-foot-thick walls with rubble between the stones.

14. The most interesting buildings were the completed ones -- made of mud brick. Like Fathy's project, no formwork was used. And as he said in his book, the designer had to work in human scale because of the inherent quality of the material. Mud simply won't span large spaces.

Village of Harranta

This tiny "village of artists" continues to haunt my memory. It's so fresh. So logical. So unpretentious. With so many beautiful forms and spaces. I had never seen anything quite like it—a masterpiece in mud brick masonry.

The buildings make sense. Many strong qualities exist: elegant simplicity, artistry, warmth, humility, dignity, structural logic, forms and spaces indigenous to the region, humble materials, proficient use of natural lighting and natural ventilation and great sensitivity to the needs of people—these people who make their living as artists.

Going through Harranta I found myself asking: Why can't urban forms and spaces be as kind to the users as the forms and spaces in this little village are? Why can't city buildings respond to the needs of the people, the climate and the region as these rural buildings do? Most of the recent buildings I saw in Egypt I could have seen in Kiev, Russia. Or in Tulsa, Oklahoma. Not the buildings in this village. They were unique—unique to Egypt.

I hope these photographs by Jim Haecker convey to you the architectural spirit of Harranta.

Nabil Gali, righ

15. The spaces were marvelous. Made me feel very important. That exhibit down at the Smithsonian will never find a better place for hanging the tapestry.

16. I thought of Louis Kahn, that great designer who died last year, when I saw these mud buildings. The forms were so fresh, so logical. And those wonderful spaces with the dome vaults and barrel arches.

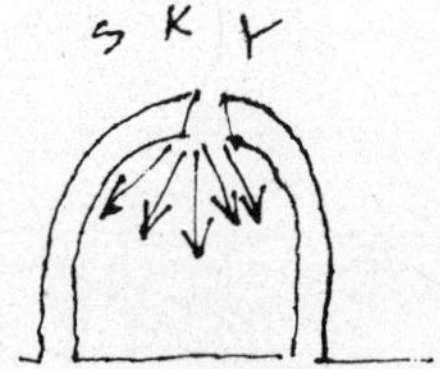

17. And, like Kahn, Architect Wassef was a master at daylighting. The museum required no electric lighting. Even the notches for displaying small sculpture had their own daylight spotlighting.

18. The greatest building of the group was for the chickens and ducks. Pure beauty. Great strength in the logic of form. Sculptural, aesthetically correct in every way, yet with absolute economy of means. Where mud was needed, there was mud. Where mud was not needed for strength, there was no mud.

19. I'm probably nuts, but this village impressed me as much with its simple beauty and

sincerity as did the great pyramid which was in seeing range. Architect Wassef helped people to live in a better world. I noticed two of the artists were physically handicapped. Where else in Egypt could they make a better living and live in a better environment?

20. Aleen told you about visiting the Galis' apartment, but I took her pencil away before she had time to tell you about Nabil Gali's wife, Azza Fahmy. Note she did not take her husband's name. Like Aleen didn't want to, but I made her.

21. Azza Fahmy is the only woman silversmith in Egypt. We bought a ring for my daughter, Susan, in dedication to the feminist movement. Azza is good, damn good. Her designs recall Egypt's history but are fresh and bold. I bought another piece for my beautiful spouse.

22. Goodnight. What a wonderful day it's been. Glad Aleen shared it with me. Tomorrow we go our separate ways. She goes to Alexandria with Margie Bush and her friend, Martha, and I go to Suez with the task force.

23 January 75 - Suez, Egypt

1. Today was the day I've dreaded -- war-damaged Suez. No doubt about it, I'm not a warrior.

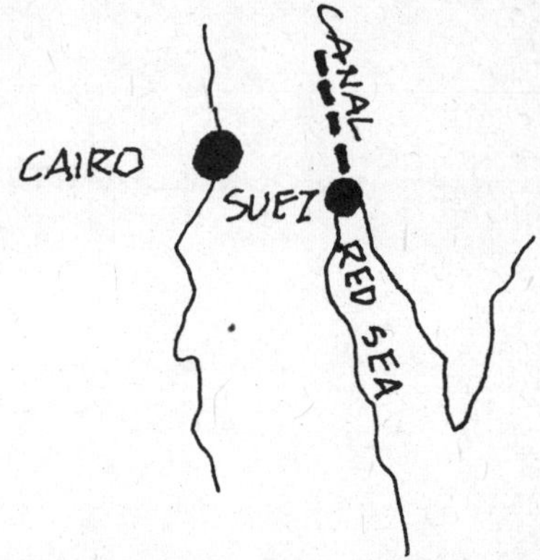

2. About 15 of us, Egyptians and Americans on special permit, motored to Suez from Cairo in a very nice bus. Went through miles of military area and scores of carefully guarded gates. Had many interesting discussions with our Egyptian colleagues, adding to my general knowledge about the building situation here in Egypt and more specifically in Suez.

3. Permit me to ramble a bit, as if I have not been doing it already.

4. The city of Suez took a hell of a shelling and invasion during the June 1967 war. Back in Houston, I read an article in the 21 December Houston Post, "Egypt Pushes Suez Rebuilding," which stated that in six and a half years of warfare, Suez City suffered 80 per cent destruction by bombs, rockets, artillery and small-arms fire. All but a few of the 264,000 persons who lived there left for safer ground. Nearby Port Tewfik was destroyed and evacuated. Port

Said and its suburbs on the Mediterranean Sea lost 20 to 30 per cent of their buildings (310,000 persons lived there before the war).

HE'S ON OUR COMMITTEE → 5. The article quotes Ismail Osman, technical director of Egypt's housing, as saying that so far more than $500 million has been sunk into the ambitious task of restoring the cities of Port Said, Ismailia and Suez to the state they were in before the war. Millions more will be spent in the coming months, according to Osman.

6. Egypt is trying to rebuild bigger and better at the entrance of the city of Suez with 3,200 new family units. I've heard they are being built for the middle class, but are low in cost. Completion date is August '75.

7. Seemingly the most reliable figures, allowing for the confusion caused by language difficulties as well as guesswork, are:

a) 1,600 family units will be completed by March. At least that's the target date.
b) 1,600 more family units will be completed by July.
c) The ancillary structures will be completed by

August.
There are nine general contracting companies doing the work; each has a certain area.

e) Workers use a nine-hour day, but most often go over and up to 12 hours.
f) There are <u>11,000 workers</u> on the site.

Dig that last figure? The lack of mechanization causes this.

8. Similarly scaled projects are going on in Ismailia and Port Said.

9. It's a good thing I'm an architect. I would have made a lousy doctor (blood scares me, particularly mine); I would have made an awful lawyer (no patience for untangling legal red tape and no talent for searching for legal loopholes). But I would have made an even worse military man. Destroying buildings makes me sick. I don't even like to see the old, obsolete arks destroyed. War is waste. I saw this destruction in Germany, Russia and Japan and I'm not looking forward to seeing the bomb and shell damage in Egypt. My life has been trying to make this world a little better place through giving it more functional and beautiful build-

THE ARAB: COURTEOUS TO A FAULT, SPURRED BY A VESTIGIAL LAND-HUNGER AS OLD AS HIS RACE, AN IMAGINATION AS FIERY AS THE RAW DESERT SUN, AN ANCIENT FOLK MEMORY SCARRED BY A SENSE OF HISTORIC INJUSTICE AND A PRIDEFUL SENSE OF IDENTITY ANCHORED IN THE KORAN AND THE VISIONS OF THE PROPHET MOHAMMED.

JERRY M. LANDAY
HOUSTON CHRONICLE
16 MAR 75

ings. I don't want to change now, because I know how important buildings are to happy people. It's great to be a builder and not a destroyer.

10. The Houston Post article quoted Osman as saying that 980,000 persons now live in the major population centers, 112,000 less than in June 1967.

"Egypt does not intend to stop with the mere restoration of the area. Long-range plans, overseen by a panel of international consultants, call for major improvements in the area and introduction of elaborate telecommunication and transportation networks, tax-free working zones, a stable infrastructure and all the amenities that would attract the world's industrial giants to set up shop there over the next 25 years."

11. Ismail Osman says, "This area has special potential. You don't have the good weather elsewhere like we have the year round. You don't have all the cheap and efficient labor, and I think the geographic location is tremendous."

ISMAIL OSMAN MADE A GREAT CONTRIBUTION DURING OUR WORK SESSIONS.

12. Housing and Reconstruction Minister Osman Ahmed Osman, the builder of the Aswan High Dam and overseer of the canal building projects, predicts it will take five more years and cost Egypt $5.9 billion to raise the area completely from the ashes of war, change parts of it into garden tourist centers and turn other sectors into industrial zones.

13. Enroute to Suez I saw quite a few tent villages. Having always had an interest in wind control, I was impressed by the method of digging down to nearly head height, then staking out the tent, and using the piled-up earth as berms to create a wind shadow, so as to protect the tent from the strong winds. Small buildings were done that way, too.

14. I also got more, but conflicting, information on expanded lime/sand blocks. Instead of its strength being 2/3 Nile silt bricks, it's 1/3 the strength. I bet that's closer.

15. Our materials experts from Egypt have set their hopes on making bricks from the many deposits of shale clay which can be broken down to soft clay. They say it can be made as strong as

the Nile silt bricks and probably as cheaply and more accurately if put into mass production. Ah, hope.

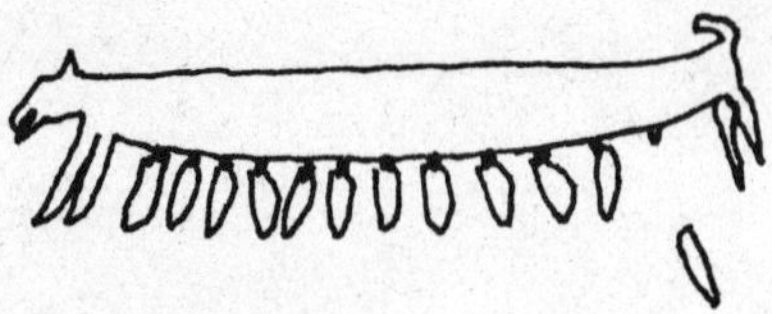

16. For you feminists, I saw no women workers like I did in Russia, which reminds me of what Aleen remarked when she heard Azza say that her dog had 12 puppies: "Oh, that poor dog!" That's empathy?

17. The new housing project is a great tribute to Egypt's managerial skill of getting things done. What they are accomplishing in putting "a roof over their heads" is quite remarkable.

18. I have an "on the other hand," however. On the other hand, the buildings as places for people will not be successful, I am sure. I shouldn't prejudge the project, but there was sufficient structure up to make me sick. The 10-year-old, 20-year-old and, for sure, 40-year-old buildings were much better. The new ones are just not people places. Too foreboding. Too nervous, looking as if there was about to be another invasion. Too raw in textures and color. No human scale like the mud brick projects I saw yesterday. No surprise vistas. No

variation in heights -- all five stories. Too jail-like. Too monotonous. Hopefully landscaping and people circulation areas will give some human values. JUST PLAIN DREARY.

19. They had completed one apartment and furnished it for demonstration purposes. Wasn't impressed with details, layout or furnishings.

20. From an architectural point of view, what I saw was bad design. If second-year students at Rice had designed the buildings, we would have flunked them.

21. I wanted Professor Shafic to be with us -- to get his opinion. He should recognize good design. He had to leave our group to serve as a juror for an international competition in another country. I bet he shares my views, but I better not try to speak for him. He is most capable of speaking for himself.

22. The buildings were load-bearing stone walls (I believe there was a first floor with four stories above. Just too much to absorb). Floors and roof were concrete.

23. Continuous footing was of brick (Nile silt brick) and ranged from 1 1/2 meters in depth to 6 meters. Because of varying water tables, brick footings were waterproofed.

24. The stone walls were like peanut brittle, shaped at the wall site by masons to fit with minimum mortar fillings. I think the rawness of the stone added strength to the looks perhaps, but no luster or charm to the project. Yet stone laid at random can be very beautiful if laid right.

25. Concrete for prefabricated floor tile (terrazo), roof decks and window sills was mixed by laborers (no mixer) but the manufacturing of these units was done very efficiently with some simple, but basic, equipment. I was most impressed that the contractor put little factories on the site for special items. The tile was put under hydraulic pressure (like Mexican tile).

26. Smooth reinforcing bars were generally bent during haul to the site because of poor transportation equipment. Understand the steel suffers a 5 to 10 per cent loss on handling.

27. Some of our group questioned the waste in not using deformed reinforcing bars. One bright young Egyptian civil engineer to whom I took a shine because of his wit, knowledge and intelligence, said, "With light loads like we have, the savings would be meaningless." I'll not argue.

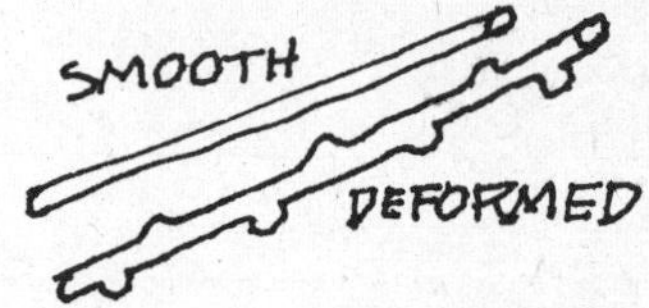

28. The bright young guy is Tarek Selim, with the Ministry of Housing and Reconstruction. He was a student of our Egyptian coordinator, Dr. Ahmed El-Erian.

29. The most beautiful architectural form (I'm sure an engineer must have done it) was the curvilinear brick walls around large storage tanks for gasoline near a refinery. They were put there for protection. After seeing what shells did in Suez, I can hardly believe there is any protection with a brick wall. I remarked, "That protection must be psychological -- for the tanks." My friend, Tarek, thought that was very funny. No wonder I like him.

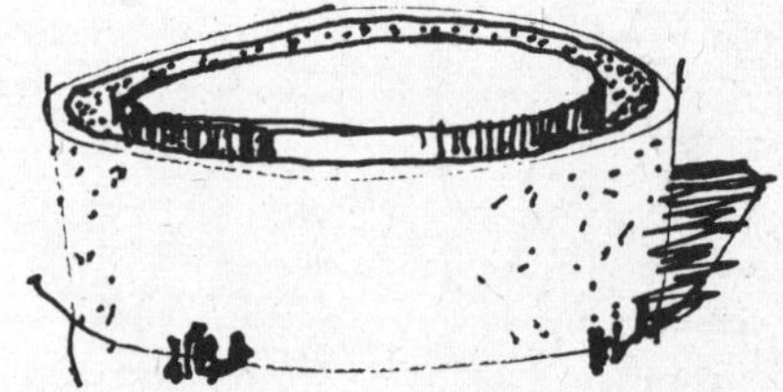

30. The Suez I saw was a town of males. Counted only five women during the morning (a task I am sure my daughter would have wanted me to perform).

31. Driving through Suez I saw a couple of kids do one-upmanship on all those people in Cairo hanging on the backs of buses, streetcars and electric interurbans. This time they were hanging on the back of a train. Jim Haecker remarked, "Give those kids paper sacks and they would stick them over their heads."

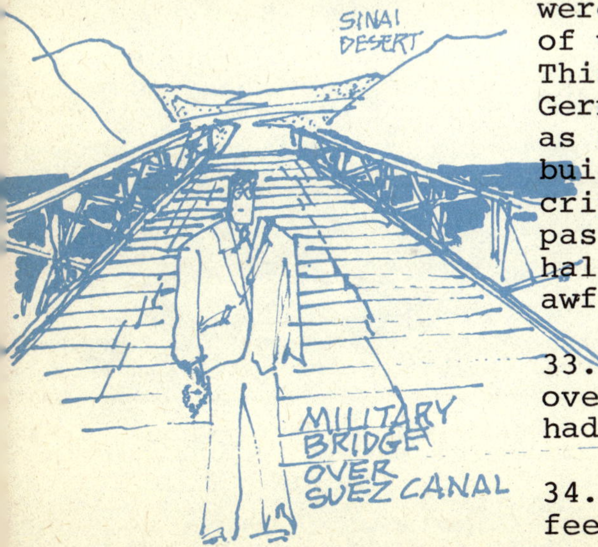

32. The results of shelling the city of Suez were effective, I suppose, from a military point of view, but awful and devastating from mine. This shelling was different from what I saw in Germany, Russia and Japan. No blockbusters as such. No great craters. But nearly every building looked like it was shot through and crippled. Looked like some kind of plague had passed through, leaving only the carcasses, half-eaten by vultures and decayed. For me an awful sight.

33. Our hosts took us across a military bridge over the Suez Canal to see where the Israelis had their guns.

34. We walked across the bridge, about 200 feet. A military bridge is what the Egyptian troops used to regain lost territory. The

water was beautiful, blue and clear. And back on the bus we rode through the Sinai Desert to the gun emplacements.

35. As in Suez, we continue to see grim reminders of death and destruction. Destroyed tanks and equipment, even some seemingly marked graves. Land mines had been cleared from the area, but I felt more secure when our driver drove down the middle of the road.

36. The gun emplacements were impossible to see until we drove behind them. The gun wells were protected with great steel nets of stones laid over steel rails and connected with winding trenches.

37. The guns were all destroyed when Egyptian foot soldiers stormed the embankment, but we could see how they had looked.

38. I knew the answer but I asked the question anyway to get it out in the open: "What kind of guns are they?" My young Egyptian friend, Tarek Selim, replied, "I'm sorry to tell you this, but they are American -- 155M, about 6 inches."

THE CLOSING OF THE **SUEZ CANAL** IN 1967 COST EGYPT $2 BILLION IN LOST REVENUE. SINCE THEN THE COUNTRY HAS DEPENDED LARGELY ON TOURISM AND AGRICULTURAL EXPORTS FOR INCOME. EGYPT HAS ALSO RECEIVED SUBSIDIES FROM ARAB ALLIES (NOTABLY SAUDI ARABIA AND KUWAIT) AND SUBSTANTIAL AID (MOSTLY MILITARY) FROM THE SOVIET UNION.

39. Imagine the hell those six guns did. Only six guns. They completely destroyed a beautiful port town of 264,000, no less beautiful than Corpus Christi or San Diego. It was systematic destruction -- each gun aimed to destroy a certain section of the city.

40. I guess history will have to be the final judge as to whether the Egyptian armies really won a military victory in the war. There's no doubt, however, that the outcome restored confidence and self-esteem in every Egyptian. And no doubt Egypt's President Anwar Sadat was boosted as a prestigious leader of the Arab world.

41. My son Bill, who's my best source for reading the political climate -- local, national or international -- found an interesting article on Egypt after the war in Time Magazine, 20 May 74. The article quotes Sadat as saying, "For five centuries, our people and our Arab nation have lived in the cradle of defeat. Victory processions roamed the streets, but they were the processions of foreign invading armies, not ours. They hoisted the banners of victory in the squares, but they were the banners of occu-

pying countries and not ours. Today I declare to our people and to the whole Arab nation that the centuries of backwardness and defeat are past and gone." Sadat made this speech at a parliamentary victory celebration in Cairo after the October war.

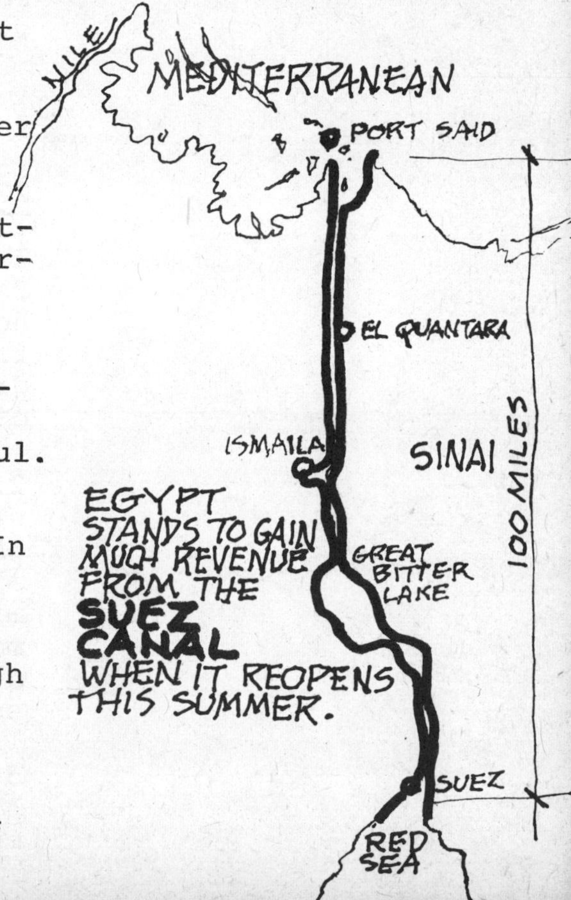

42. As for the use of guns, I have the aesthetic curse. Frankly, I wouldn't have minded hearing guns blasting the hell out of an empty housing project like the one we saw, just for the target practice. Maybe the results would be to provide the people with better design resulting in beautiful, livable places. Some of the destroyed old buildings were still beautiful.

43. My barbs at the bad design should be directed, I'm told, to the Arab Design Office. In 1961 this was the largest private firm. Then came nationalization. I wonder how good they were then. I'm sure there is some good reason for the sophomoric designs. Probably not enough time. But good designers don't need as much time as immature, inexperienced designers.

44. This day was just too much. I'll lose sleep over it. Surely no damn good can come of

war. Or can it? Is the U.S. better because of our Civil War?

23 January 75 - Trip to Alexandria - Aleen

1. Margie Bush, her friend Martha and I left Cairo early in the morning by car for the 2½ hour drive to Alexandria. Carts piled high with fresh produce were coming into the city as we drove out. They were loaded with carrots, huge cauliflowers and cabbage.

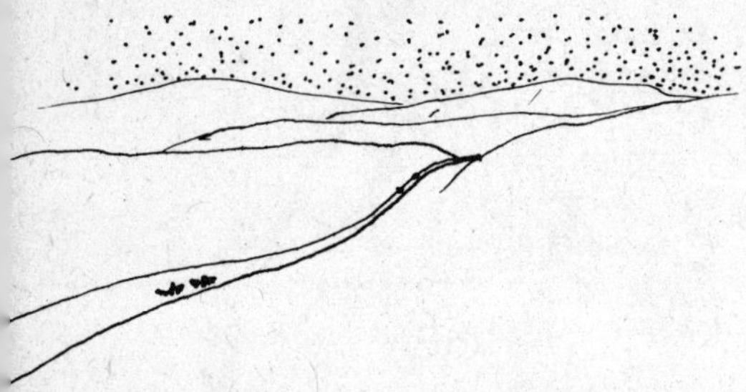

2. Took the desert highway. The driver asked us not to take pictures for security reasons since there were some military installations along this road. We drove through miles and miles of pale yellow sand -- just like in the movies. As we neared Alexandria, the desert was dotted with clumps of green because they do get some rain near the sea coast.

3. Egyptians grow ecstatic when they speak of Alexandria. After seeing dusty and crowded Cairo, I understand why. Alexandria was washed

clean by an early morning shower, there was no bumper to bumper traffic and the Mediterranean was a heavenly color.

4. Saw the palace where King Farouk was living at the time he was deposed and had to leave the country. Know he and his family were sorry to leave their 85-rooms-plus summer home that overlooked the sea.

5. Had lunch at the Santa Lucia, a very nice restaurant, where I had the biggest shrimp I've even seen -- some were fried and some grilled. They were twice as big as our jumbo Gulf shrimp and very delicious. Had a nice white Egyptian wine with lunch and finished it off with demitasse cups of dark, sweet Turkish coffee. Brass finger bowls filled with warm water were brought to the table when we had finished.

ALEEN, SURELY NOT!

NOT BIGGER THAN OUR **TEXAS** SHRIMP.

6. Came home by what the driver called the Irrigation Highway. Here were lush green fields (irrigated) where they harvest four crops a year. A hub cap came off the car as we were passing one of the fields and a whole family came out to help the driver find it. They hoped to get a few piasters for their efforts. One of the workers

did find the cap in the alfalfa field but the driver ignored him. The man made some angry noises and gestures but the driver was not the least bit perturbed.

7. Passed many adobe houses with straw roofs. Sitting on top of a number of them were mud beehive shaped objects. I learned these were pigeon houses. The farmers raise pigeons to eat and sell. Now I remember that pigeon was on the menu at the hotel.

8. At dusk we passed the workers leaving the fields. Most were walking, but when they had a burro, the husband would be riding and the wife walking along behind him.

9. Margie Bush told me about seeing a couple sitting on the floor at the airport with a bundle of their possessions along side them. As they got up to leave, the man put the bundle on the wife's head. He put on his coat and they were off. The man was empty-handed. The word on women's lib hasn't reached this strata of their strata. Back to Bill.

24 January 75 - Cairo

1. Reading the papers in Cairo makes one see the Arab side -- right or wrong. Here's an excerpt from this morning's The Egyptian Gazette's main editorial, entitled "Battle for U.S. Opinion":

"The majority of Americans are against military intervention should the Arabs impose another oil embargo on the U.S., according to a Gallup poll published yesterday in The Washington Post. This may show that the effect of Zionist propaganda in the U.S. is not quite as strong as it was because there can be no doubt that the Zionists would favor any action against the Arabs, including military intervention. This does not mean, however, that U.S. opinion is now to a large extent pro-Arab. But it does mean that the Zionists do not have it quite so much their own way as before.

"The turning point was the October war and its aftermath. The Arabs' fight for their freedom came as an eye-opener to the majority of Americans. It brought home the truth that the Arabs had a cause for which they had to go to war and that cause, if not vindicated, might involve the world in a conflagration."

The Egyptians are happy about the Gallup poll.

2. Our U.S. group met with Bill Templeton and Dan Pfouz at the American Embassy. Templeton warned us to make sure we distinguished between problems and programs -- that our job is only problems at this time -- helping the Egyptians to identify them.

3. Next we went to Cairo University (Enrollment is somewhere between 84,000 and 105,000. Heard both figures.) for a work session of the joint study group.

4. Neils Thompson led off the discussion by presenting 10 salient considerations. He listed: a) Determine the production of construction materials as to quality/quantity and per cent of annual need; b) Determine which construction materials are most needed (priority) and determine which steps should be taken to adjust existing operations or plan new operations. Locate plants; c) Take each industry (plant by plant) and analyze what are the most economical steps needed to upgrade production, quality, efficiency

and program each over a 10-year period; d) Determine the labor-saving devices that would increase quality/quantity of production that can be most easily introduced into building construction; e) Introduce selected, simple prefabricated systems; f) Analyze transportation methods. Find ways for improvement. Consider transportation of materials; g) Establish technician and skilled labor training program; h) Make an analysis of building codes, building inspection construction safety and enforcement; i) Determine the availability of the latest technical information; j) Determine the critical need for laboratory equipment.

5. Phil Bush followed next with his list of four: a) What is the total organization structure, responsibility and authority for each step in the building process from exploration to finished building? Explanation of social side effects; b) "Location" trade-offs for production plants (Refer to Thompson No.6); c) "Type of plant" trade-offs; d) Massive housing (high-rise) vs community patio.

6. Then Bill Cullen made his contribution: a) Anticipate getting into petrochemicals; b) Look

seriously at manufacturing gypsum board; c) Develop paints and finishes; d) Study building service systems; e) Much concern with windows and doors. Cement, concrete, brick research okay; f) Building process - CPM.

7. When it was my turn, I told them I had no salient points and instead gave them a sermon on humanization of buildings -- that people were more important than bricks.

HERE'S FATHY AGAIN.

8. Ibrahim Gafar, Head of the Structural Engineering Dept. at Cairo University, apparently liked what I said and told us about Hassan Fathy. He likes Fathy's thinking and his plans for teaching workers to build a village; then having them move on to build another village. It seems you are either for Fathy or against him.

9. Prof. Gafar gave this account of Fathy's project. The houses in a new village had white interiors. When the people moved into their new homes, they didn't like them. Too much alike. Not in tune with their culture. The dwelling units were not individualized. Too monotonous. What did the users do? They put mud all over the walls. Some inserted ceramic plates into the wet

mud plaster to create an environment of their aesthetics.

10. That's what I tried to tell our joint group. No matter how advanced the technology, how fast construction goes, what the price tag is, or how beautiful the buildings seem to the designers and builders, if, when the people move in, they don't like the space and form, then the project is unsuccessful. When they like it, then there is architecture.

THIS IS WHAT **ARCHITECTURE** IS ALL ABOUT.

11. Jim Haecker continues to do a good job of leading our group and serving as our diplomat. He's good at it, primarily because he loves Egypt and its people. During our work session he urged cooperation among the various governmental agencies related to the building agencies. Like ours in the U.S., there were jealous fences that needed removing.

12. A member of the Egyptian group who was most impressive with his knowledge of the building industry is Ismail Osman with the Ministry of Housing and Reconstruction. He's quoted in the Houston Post article I talked about earlier. He told the joint committee that Egypt was committed

to concentrate its building program on the Suez Canal area. Reason? 800,000 people lived there before the '67 war. Following the war, 20,000 remained in the area. Now about 500,000 have returned.

13. Ismail Osman, who, as I understand, is the nephew of the Minister of Housing and Reconstruction, went on to say that Egypt needs to build 100,000 housing units every year. It builds 15,000 units. That's why our group is here.

14. The Minister of Housing and Reconstruction doesn't mind asking for outside help -- another reason we are here. They have hired a few U.S. firms to advise, do feasibility studies, planning and design. And other countries too.

15. Another deep-thinking Arab in our group is Hassan Kamel Imbabi, Professor of Engineering and Head of Building Materials Dept., National Research Centre, Cairo.

16. My architect friend, Dr. Hussein A. Saleh, submitted an eight-page report for his and Professor Yousef Shafic's contribution. Very com-

plete with broad scope.

17. Then we had the very personable Dr. Mohamed Ramez, of the Building Research Center, who called for action based on planning, material and design. He was rather tired of all the talk and asked, "How can we get something concrete out of our meetings?" Rather odd coming from someone whose specialty is clay brick. Don't hit me.

24 January 75 - Cairo - Evening

1. This afternoon we went off to see Mr. Big -- Osman Ahmed Osman, Minister of Housing and Reconstruction -- at his farm on the outskirts of Cairo. As I said earlier, Osman was the principal contractor on the Aswan High Dam.

2. The Minister was waiting for us on the terrace. Most impressive, in a blue sport coat outfit with yellow "worry beads" in his hand. Aleen asked me, "What quality?" Damn good. With his responsibilities they had to be.

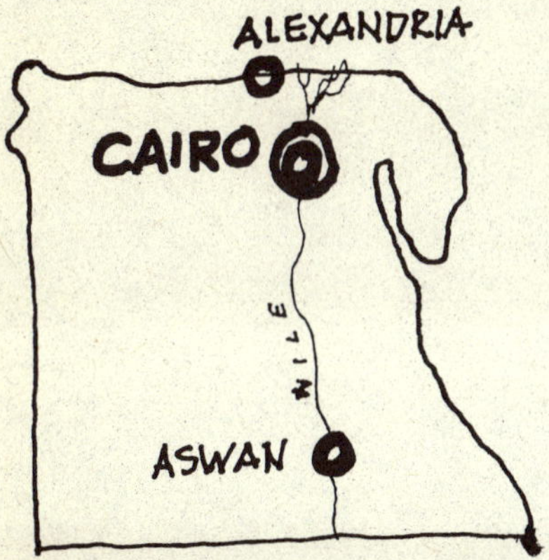

PRECIPITATION: ANNUAL
ALEXANDRIA 7"/YR
CAIRO 1+"/YR
ASWAN 0 - TRACE
FROM THE MEDITERRANEAN COAST TO THE INTERIOR RAINFALL DROPS OFF SHARPLY.

TEMPERATURE: JULY
ALEXANDRIA 85° AVG. MAX.
CAIRO 96°
ASWAN 106°

3. The Minister was most cordial. He was knowledgeable. And most open. Acted as a host, not as a judge of what we were doing.

4. He's a thinker. But more, he's a doer. Abhors government red tape and pessimism. He's very proud of being referred to as a "man of action." Should be. He's getting things done. And fast.

5. His first assignment from President Sadat concerns a $6 billion reconstruction of the Canal Zone, including the cities of Suez, Port Said and Ismailia.

6. Osman's also concerned with reclamation of land on both sides of the Canal. A long-range goal is to build a system of concrete culverts beneath the Suez Canal to carry water from the Nile to irrigate the Sinai desert. Time Magazine (20 May 74) reports that "at least 350,000 acres of wilderness can be reclaimed by this process. 'We wouldn't have done this before the building of the High Dam,' says Osman. 'If there had been a low Nile, we might have siphoned water over to Sinai. But now we have a guaranteed supply of water.'"

7. Osman has other dreams -- to reclaim 800,000 acres of land below the dam and the reclamation of nearly 2 million acres of the Western Desert. "Go West, young man." And he's only 57.

8. Electric power from Aswan will be sent to the desert and used to pump up the water and irrigate the land. Within the next five years, 4,000 Egyptian villages will have electricity for the first time.

9. According to Time, "in a test project 100,000 transplanted Egyptians are now living in the Kharga Oasis at the southeastern end of the desert where they successfully raise crops and livestock. One farmer, Mohammed Mahmud, happily told Time correspondent Wilton Wynn: 'I used to live in a village in the Nile valley where I was able to rent only two acres of land. I moved over here and now I own six acres of land, two cows, four goats and a donkey. In the old village there were 17,000 people and only 1,600 acres of land. Over here, there is enough land for everybody.'"

10. My daughter Susan would be interested in knowing, if she does not already know, that

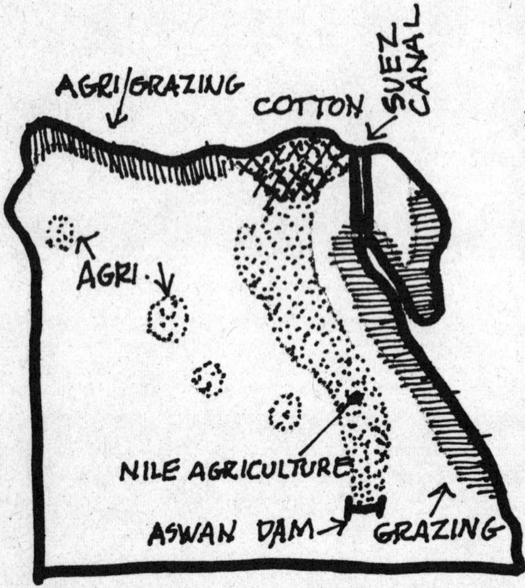

Egypt's leaders are fighting to limit families. For example, Mohammed Mahmud, mentioned above, now has 11 children. And a pregnant wife. With support from progressive Islamic leaders, the government has set up birth control clinics. A monthly supply of pills, by the way, can be bought for about 12 cents. Already there is a slight drop in the birth rate, but this is probably due more to Aswan's electrical supply -- more lights and television. One social worker said, "They have other ways to spend time now."

11. Osman's farm, about 60 acres, is a very pleasant one. New construction everywhere as if a builder lives there. One does, because before the nationalization of the building industry, he was Egypt's largest private contractor. On his place were birds and animals. Two gazelles were grazing on his front lawn.

BEFORE ASWAN HIGH DAM THE AREA OF CULTIVATED LAND AMOUNTED TO ABOUT 3% OF EGYPT'S 386,000 SQ. MILES.

12. Osman loves cut stone. Had 10 or 12 free-standing stone arches through which approaching cars drive. Sheds were made of stone barrel arches.

AGRICULTURE IS THE LARGEST SECTOR OF THE ECONOMY.

13. He said that he insisted that the new city at Suez be traditional construction -- load-

bearing stone walls. Trouble at first finding enough stone masons. Lots of stone. No problem there. Did take long to train the masons.

14. The paradox of his choice in using stone, if there is one, is that a load-bearing stone wall, with each piece shaped on the job site, is about the slowest way you can put up a building. And in our country the most expensive. Years ago CRS was hired by Duke University to "invent a new Gothic that we can afford" and Duke had its own quarries. CRS used prefabricated panels of stone to save labor cost.

15. Osman apparently had all the labor needed -- 11,000 men on one job site. I restrained myself from saying that load-bearing stone wall systems were the slowest of any system. How could I? You can't argue with success. 15,000 completed housing units in six months! That's success. Osman's built stone barrel vaults up to four meters.

16. Osman said his next concern is mass-production at the quarry site. To get cut stone out accurately and fast. He's willing to pay any foreign firm to help them do this.

17. The Minister emphasized over and over, "We want to develop this country. We cannot develop it alone."

18. I wonder how many countries are helping Egypt. Poland seems to have furnished a lot of machinery and equipment.

19. Osman really is a man of action. Our colleague, Dr. Mohamed Ramez, said that in his research labs he had developed a new form of clay shale and it was ready for production but he couldn't get anyone of high authority to listen to him. Osman immediately responded, "Give me the data and I'll see that we'll get that brick in buildings fast." I believe he will. Ramez was overjoyed. He had been working on the research project for four years. Had written reams of letters and reports "to get someone to listen." Someone did.

GREAT VISIT WITH OSMAN

20. The conference with the Minister of Housing and Reconstruction in the sun on the terrace of his farm was unquestionably the highlight of the week's working session. Osman exuded enthusiasm. Our Arab friends on our committee were as pleased as we. Of course Ramez was so pleased with the

whole affair he nearly cracked his face from smiling. A great morning!

21. A note to the ladies: We were served tea, Egypt style. In tiny demitasse gold cups with the handle curled up near the bottom. The design forced my little finger to point up. I looked like Mrs. Gotrocks at a social tea. The tea was good with a strong peppermint taste. Only trouble is the silly little cup just held a tablespoon full.

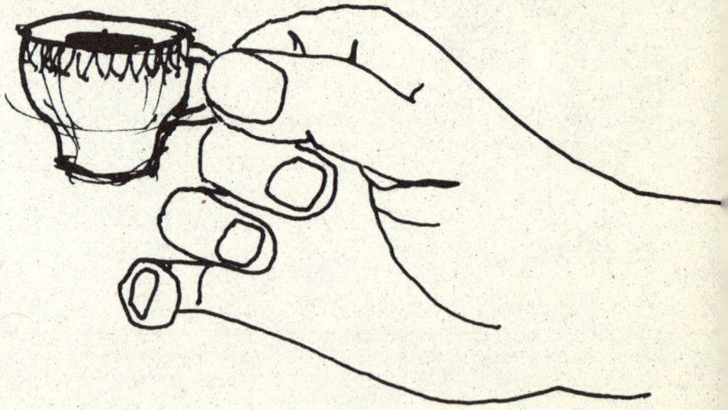

22. After we had spent nearly an hour with the Minister, he diplomatically signaled to his nephew that our time was up. We said our goodbyes, stood for the usual pictures and were on our way to see the rest of his farm.

23. Everything was cut stone in the tiny village on his farm which included houses for his farmers, a school and a mosque. He generously gave this to his people out of his own pocket, we were told.

24. The buildings were all new or under construction. None being used. Considerably better than the mud huts we saw earlier.

SCHOOL

25. The mosque was most impressive with highly complicated plaster panels -- manufactured at the site -- for interior linings of walls, columns and altar screens. The dome and roof structure was of heavy reinforced concrete.

26. Let's go back to Minister Osman during our conference on his terrace. "Socialism," he said, "means you should have a house for your man, a hospital for your man, a school for your man, a park for your man, etc." He's just about provided all of this on his own 60-acre farm.

27. The little school contained five classrooms, each a barrel vault in width. I'm sure structure determined the size, not education. The ceiling was a disappointment. The vault had been plastered, then scored to look like stone. Mr. Osman could have used a more innovative and sensitive designer.

28. The little houses, also with stone load-bearing walls and stone vaults, suffer from poor planning. There was a great opportunity for the "dog trot" idea and spacious outdoor living. We cannot afford that beautiful stone construction in our country for low-cost housing and school

envelopes. It emanates permanency and structural quality. But not human quality. Wouldn't it have been great if Mr. Osman had had highly creative architects to help him fulfill his dream for his people -- not only on his farm but on his giant housing projects, too.

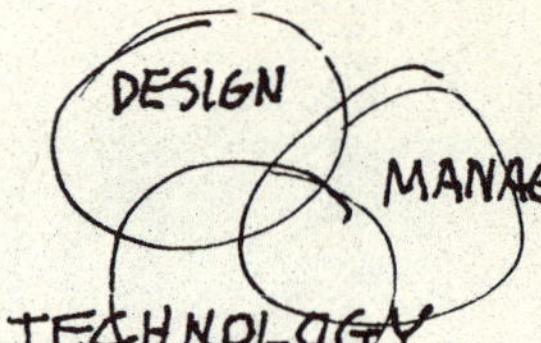

29. Egypt has a custom I like. Since the time of the Pharaohs, the ibis bird has been considered sacred. Never to be killed.

30. I wish Egypt, and our country too, considered architecture that precious. People must love their buildings to be happy. Or love their caves. We architects must realize this. If mud walls are necessary to their happiness, then mud walls it should be. If "letting the man," as Osman would put it, build his own personal environment and thereby achieve his own satisfaction, then we should let the man do it. Architecture is a personal experience. Forms and spaces must fulfill certain 1 physical, 2 emotional and 3 intellectual needs for that person. If they don't, then there is no architecture. The person -- that sacred user -- is the final judge of whether or not the forms and spaces possess architecture.

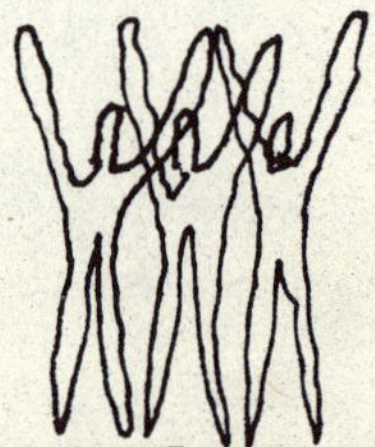

31. From the "farm" -- quite a farm -- our group of 12 drove back into town. Ismail Osman had arranged for a luncheon boatride on the Nile. Beautiful day. Smooth water except for a speed-boat that ran circles around us. A magnificent view of the city. Great trip!

32. I was rather amused to see clothes drying from balconies of luxurious apartments over-looking the Nile. I believe I would trade a view for a washer anytime. But I don't wash clothes.

33. Then back to work at Cairo University where our joint study group devised the format and agreed on the content in general. A remark-able feat. The chemistry was there.

34. A great day -- most productive.

24 January 75 - Cairo - Aleen

1. Was on my own today and decided to see where the natives shopped. In this large shopping area

I never did see a department store as such, but I have never seen so many fabric stores.

ALEEN IS BACK.

2. There must be many good tailors and seamstresses in Cairo because I saw thousands of bolts of fabric of every description. There were many elegant ones with gold and silver threads and all kinds of printed cottons and silks.

3. One specialty shop had nothing but buttons, laces and trims. It was doing a booming business, even though this is the Moslem Sabbath (Friday).

4. At noon the call to prayer came over loud speakers in the business district. I saw two men come to this public arcade where straw mats were rolled up and leaning against the wall. They each unrolled a mat, took off their shoes and knelt on the mat, touching their foreheads to it.

PRACTICING **MUSLIMS** BOW IN PRAYER (SALAT) FIVE TIMES DAILY IN THE DIRECTION OF THE HOLY CITY, **MECCA** IN SAUDI ARABIA. COLLECTIVE PRAYER IS ENCOURAGED ESPECIALLY ON **FRIDAYS** WHEN AN OFFICIAL, THE IMAN, DELIVERS THE WEEKLY SERMON.

5. I saw no women praying and women are generally not admitted to the mosques. I understand there are two mosques which have balconies where women may go. As far as I know, women can ride in the front of the bus. ☺

THE **PAPYRUS** PRESERVED RECORDS OF EGYPT FOR THOUSANDS OF YEARS. LITTLE WOULD HAVE BEEN KNOWN OF THE **EARLY EGYPTIAN CIVILIZATION** WERE IT NOT FOR THIS DURABLE WATER PLANT.

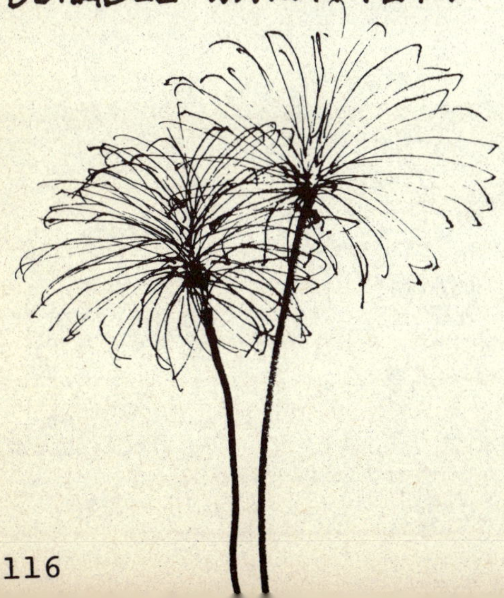

6. At the Papyrus Institute, I learned an Englishman has been working for some time to perfect the art of making parchment from papyrus as the ancient Egyptians did. Saw the papyrus cane cut into strips. Each strip is hammered to eliminate some of the juices and then soaked in four different baths of water. Each soaking takes from 24 to 48 hours. It is then laid criss-cross between linen and put in a press to squeeze out the water. This is done several times, so the squeezing and drying process takes many days. But it comes out as strong, tough parchment that can take type, paint, etc. Only the kings and queens could have afforded it originally. They used it to record how great and noble they were.

7. Bill is off the Lomotil but he has overreacted. We ate tonight at the coffee shop in the Hilton which is about four blocks from here. Never thought I'd be so happy to see a Hilton hotel, but their food is saving my life. I lose my appetite when I walk into the Shepeard's dining room. This is "Amen from Aleen." Bill takes over in the a.m.

25 January 75 - Cairo

1. Today is the last day for meetings of the joint study group. It's been quite a week. Friendships between the Egyptians and the Americans began in a most cordial climate and have flourished.

2. This morning we got a poor start. We were told that our first meeting would be at the Academy and we'd be picked up at 10 a.m. As we were waiting in the lobby, one of the Egyptians came in and said we had been keeping the Egyptian Technical Advisory Committee waiting in the Board Room of the Ministry of Housing and Reconstruction. Early in the week such a meeting had been scheduled, but was cancelled by our Egyptian coordinator. Of course, we were puzzled. Off we went to the meeting with the Advisory Committee. Didn't want to cause an international incident.

3. There were 12 members waiting for us. Jim Haecker did a masterful diplomatic job of apologizing for our tardiness and blaming the language difficulty, but we knew an internal riff was the real cause of the confusion.

THE LITTLE ARK IN WHICH THE MOTHER OF **MOSES** CONCEALED HER SON IS BELIEVED TO HAVE BEEN MADE OF **PAPYRUS.**

4. We met a distinguished group of engineers, architects and industrial people led by our old friend, the midnight airport greeter, Hamed Kaddah. He introduced his group with great aplomb and dignity. Then we introduced ourselves.

5. We had the works -- hot spiced tea served in dainty glass cups and saucers.

6. Meeting with the Advisory Committee, of course, made us late for our scheduled meeting with the joint committee. One of the Egyptian members seemed put out with us and said, "You should not have gone there. That committee does not exist." Apparently their big boss had not recognized the committee. It did exist, believe me. Impressive people, including the Deputy Minister.

7. Bill Cullen philosophically summed it up, "In Egypt one needs only patience and humor."

8. Our Egyptian coordinator was not disturbed, at least he did not show it, by our trip to the Ministry of Housing and Reconstruction. In fact, by the time we got to our meeting, each of us

had been given a copy of the first draft of the report only two and a half hours after we had turned in our part. Understand some of the Egyptians worked till the early morning on the documents.

9. The first draft is very rough, but the contents are relevant.

10. We reviewed the draft with Dr. Abdel Moneim Abou-El-Azm, president of the Academy of Scientific Research and Technology. He approved the contents. Later during our luncheon, he told us about his visit to Oklahoma State University, Aleen's and my alma mater.

11. The consensus was for the two coordinators to polish the document for completion in March.

12. Had lunch at the Egyptian Car Club. When King Farouk was throwing his weight around (and there was a lot to throw), it was the Royal Car Club. He spent a lot of time there playing cards. I'm sure the place must have lost its luster after the king was ushered out of Egypt.

13. During the luncheon I happened to mention

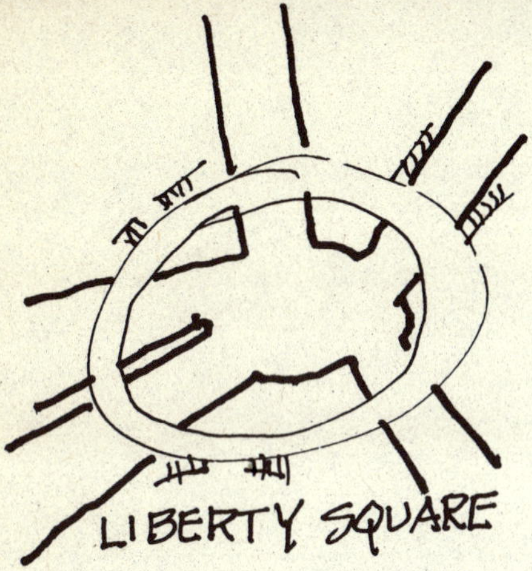

ALEEN WANTED TO KNOW WHY IT'S NOT CALLED LIBERTY CIRCLE.

that, as I understand it, King Farouk's body was allowed to be buried in Cairo, but this was done very quietly. Surprisingly some of the Egyptians didn't know this. Others confirmed it. Farouk was the last of the really big-time playboys. I always wanted to be one.

14. The lunch was a feast. Since it started four hours past my usual 11 a.m. lunch at Romana's, I was starved. I blew it. Now I'm back on Lomotil. Me and my big mouth.

15. Bill Cullen, Hussein Saleh and I walked around the heart of the city prior to lunch. When I say "around," I mean it. Cairo has a great urban circulation device on Liberty Square -- a circular raised pedestrian ramp which rises over the criss-crossing bus and automobile traffic below. Very nice, but a bit narrow. Built two years ago by Minister Osman.

16. Was delighted to see Tarek M. A. Selim, my young civil engineering friend from the Ministry of Housing and Reconstruction who went with us to Suez and served as my personal interpreter. He's a beautiful, outgoing person. Big and sharp. Reminds me of Jim Falick, one of CRS's

bright lights. When I said goodbye to Tarek, I felt very sad.

17. In fact, after the luncheon (luncheon, hell, it was a banquet) we said goodbye to all our Egyptian colleagues. I had a long goodbye with Dr. Ahmed Hassanein. Phil Bush had casually mentioned to Hassanein that his wife Margie's birthday was today and, presto, Margie had a beautiful bouquet in her room. Aleen got to appreciate the bouquet too, since Phil, Margie and their friend, Martha Beckman, were leaving very early the next morning, and Margie gave it to Aleen.

18. And speaking of flowers, Bill Cullen must have a secret admirer. He had flowers in his room, too. No card. He insisted that they were missent, but the delivery boy refused to take them back. Bill must be the greatest thing that's hit Cairo since "Little Egypt."

19. Aleen and I took Bill for his farewell dinner to the roof dining room of the Cairo Hilton. Going to miss that guy. But I sure as hell wouldn't send him flowers. Hope his wife Betty never finds out about this.

20. Back to the hotel. We met one of Neils Thompson's former students at Texas University. He is Dr. Salah B. Nosseir and he's Egypt's answer to Phil Bush -- short and sharp.

<u>26 January 75 - Cairo - Aleen</u>

THE BEAUTIFUL SPOUSE

1. Another nice sunny day. Our room overlooks the Nile, and we could see one of the bus-boats crossing. We could also see one of the steel nets put across the Nile during the '67 war to protect the city from floating mines which could easily be dropped from airplanes. They have spread the nets apart to let boats through but they are still there.

2. Our room is spacious and we have a nice bathroom -- one section for the commode and bidet, and the other for the bathtub, two lavatories and a small dressing table. They don't believe in cleaning the bathtub every day and the commode runs all the time, but we've learned to ignore that, except for the ring in the tub. Oh, for a box of Ajax!

3. We can't complain about the service. Press one button by the bed and a little man instantly appears to take our breakfast order or whatever. No. 2 button is for the maid and No. 3 is for the bellboy. Bill said when his stomach started aching last night, he panicked and wanted to push all three buttons. That would have started a riot. They'll do anything for a tip.

4. Hustlers are all over the place and the shop owners keep a close eye out for a likely customer. The other day I paused too long in front of a window because I was curious. There was nothing in it except empty, ordinary bottles of various sizes. Before I knew what had happened, the little storekeeper had ushered me inside, all the while telling me how wonderful the Egyptian perfumes were and how superior they were to the French perfumes that were diluted with alcohol. I was a little nervous because I was the only one in the store and he had closed the door. I quickly decided to buy an ounce of "Secret of the Desert" and one of "Lotus" at $1.70 an ounce. Then I relaxed because two other people came in the shop. From a big decanter, my little salesman poured the perfume through a tiny funnel into the bottles. The he rummaged furiously for

five minutes trying to find caps to fit the bottles. When he couldn't find any, he looked for other bottles with caps. He finally found those and transferred the perfume from the first bottle. Then he neatly wrapped my purchase in a piece of newspaper and sent me out, repeating how much my husband would enjoy using "Secret of the Desert" as an after shave lotion. Somehow I don't think Bill will be overjoyed with it.

26 January 75 - Cairo - Evening

1. At 1 a.m. this morning, the Lomotil demand returned. Too much lunch and too much dinner. Nearly too much for me. I felt awful.

2. Waiting in our mailbox this morning was this note from our fearless, peerless leader, "Bill and Aleen, here's some extra Lomotil (hope you don't need it). Sorry I could not say good-bye properly. Many thanks for a great job. Will call you in U.S.A.," signed Jim Haecker. And was I glad to get the Lomotil. Aleen said my supply had dwindled dangerously.

3. For the past four days I've been trying to get our return tickets straightened out. Seems we don't have confirmed reservations home. Our young, aggressive friend from the Ministry of Housing and Reconstruction, Nabil Ahmed, who is taking care of us, has been trying to unscramble the mix-up. He finally got us square with Egypt Air and we are scheduled to go next Saturday. Was Sunday, but no sweat. ← MY STOMACH IS HOMESICK.

4. Nice to have no work sessions today. Aleen and I leisurely started walking toward Liberty Square. Someone had told us about a shop selling fine jewelry, fabrics and artifacts -- that Neiman Marcus bought things there. After dodging cars on converging streets, stopping in a few tourist traps, not to mention being manhandled (nearly) by pushers, we finally arrived at the pedestrian circle which permits people to move above the traffic congestion. Relief from dodging cars. Aleen had written instructions on how to get there. By golly, we did.

5. Nice shop. Good merchandise. Understand Neils Thompson bought a table. Will take about four months to get to the U.S. We didn't see anything that we couldn't do without.

6. From the little shop we went to the Egyptian Museum. I had been prepared for this by Charles Lawrence and others; at least I thought I had. No one really can be. I suppose that's because it's going through a constructional transition, and having very few lights on increased the absurdity. Mind you, there is no place in the world, including the Metropolitan Museum and the British Museum, that has such a collection of Egyptian antiquities. And what did we find -- the appearance of a huge, dusty warehouse.

7. It's literally crammed full of priceless treasures. They are stuck every place. In corners. On top of each other. On the floor. In halls. Great pieces of sculpture. Exquisite small pieces. If Houston had just one relatively small figure, there would probably be a separate building to house it. Going through the museum was just much too much. Among the priceless treasures, we saw a large box with inlaid ivory figures. We were both shaken at seeing a figure of a man with a penis over two feet long. I have no comment.

8. Aleen has finally made contact after three days of trying to get in a tour with Mr. Moose

TELEPHONING IN CAIRO WAS QUITE FRUSTRATING AT TIMES. ONE EGYPTIAN SAID, "TELEPHONE SERVICE IS A NATIONAL DISASTER."

of Gazerine Travel Agency who is (we think) making arrangements to Luxor (Ancient Thebes) then on to Aswan where that damned dam is. Aleen finally bull's-eyed. Mr. Moose is alive and living in Cairo. He says our trip for Tuesday is set. Smiles. 

9. We've no doubt set an All-American record. Been here a week and still have not visited the pyramids. Will do that tomorrow morning. Have decided to get up at 5:30, get a cab without a guide, and wander around by ourselves watching the light change when the sun comes up.

10. 5:30 a.m. That's early. This morning one of the men "maids" on our floor pounded on our door, for hours it seemed, trying to awaken us. He did. I staggered to the door and told him to scat. "You've got the wrong door, damn it." That and my "illness" were just too much for my weak bod to take. Goodnight! (The bod is better.)

27 January 75 - Cairo

1. We did it -- got up at 5:30 a.m. Pitch black outside. Aleen and I, without a guide, planned to grab a taxi and head out to the great pyramids to see them in the changing light of the early morning sun. It was hard to believe there was a sun at that un-Allah hour.

2. We're on the eighth floor of the hotel. Punched the elevator button. Elevator cab came. Elevator cab doors started to close before we got in. Stuck my arm and foot in to open the door. Whamo! Doors, like giant clamps, grabbed my arm and foot. My strength couldn't override the door's inward thrust. Caught like a bear in a trap. Cussed like hell. Had no effect on the doors. Aleen had raised eyebrows and criss-crossed mouth. Finally pulled myself clear of the jaws of the man-eating machine. It hates me. Went on up without us. Caught it again coming down. When we got inside, I kicked the hell out of it.

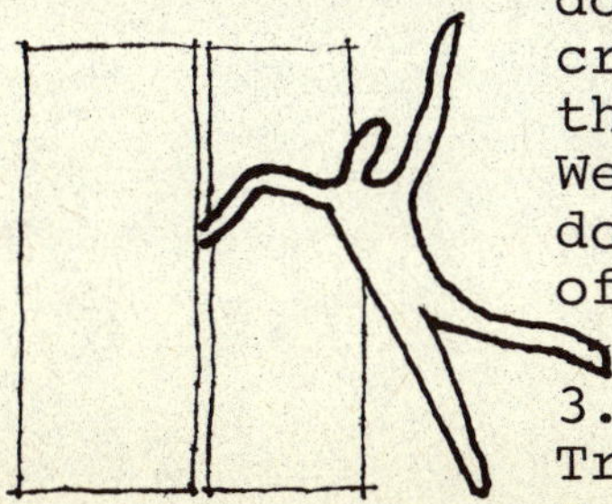

3. Hailed a taxi. Taxi driver spoke no English. Tried to tell him we wanted to go to the pyramids. No savvy. Finally he asked a passerby what we

wanted. I said we wanted to go to the pyramids. With great gestures, the man asked, "NOW?"

4. The taxi driver turned us down. Found another one who agreed with two words, "Pyramids, OK," and a big smile. Still pitch black. The guy drove like hell, racing the sun, I guess. Passed a donkey cart. He said, "Donkey." "OK, good," I answered. Suddenly he threw up both hands, still driving like crazy, and yelled with delight, "Speak English goooood." Aleen and I applauded. Cairo taxi drivers drive with great exuberance.

5. Early morning traffic was light except for donkeys pulling carts and donkey-riding men -- carrying bread, hay, onions, greens and loads of cage-like boxes.

6. Still dark when we arrived at Cheops, the largest pyramid of the big three. Asked the driver to stop and we walked up to the side of man's most lasting, if not most impressive structure. When we looked up at the side of the pyramid, the moon lit it up. Aleen exclaimed, "Bill, there's the Big Dipper." Sure enough. There it was at the apex of the triangular sides.

THE **ARAB** LOVES THE **NIGHT.** ITS COOLNESS. ITS STARRY MATHEMATICS. ITS INFINITY. I CAN EASILY UNDERSTAND THIS. EXPERIENCING THE **PYRAMIDS** AT NIGHT WAS AN EVERLASTING DRAMATIC **IMPRESSION.** BREATHTAKING -- SEEING THE BIG DIPPER APPEARING AS IF IT WERE RESTING ON TOP OF THESE HISTORICAL GIANTS.

7. Without warning, two shadowy figures appeared out of the sand and rock. Don't know where they came from. One turned out to be a guard. Just asked Aleen if he had a gun. She doesn't remember. As I recall it, he had six machine guns and 10 pistols because he scared the hell out of me. He turned out to be friendly. The other one said he was an official. That impressed us. As it turned out, however, he was an "official" guide named Frank. Just what we didn't want.

8. Aleen, Martha and Margie had gone on a side trip Thursday to Alexandria with a guide named Solomon -- a big blowhard, a namedropper who knew less than the postcards. Aleen decided Frank was one notch better than Solomon. At least he tried.

9. We then went to the next largest pyramid, Chephren. That's the one with the granite cap. Then to the smallest of the big three, Mycerinus. The thing that surprised us was that the pyramids were located on a rock hill, a jebel I believe they call it, overlooking the city. We thought they were located on the flat desert. The granite came from Aswan hundreds of miles up the Nile.

ABOUT CHEOPS AN INSERT

Before our trip, Aleen and I read an interesting piece entitled "The Great Pyramid at Gizeh" (or is it Giza? Wish they'd made up their minds -- I'm confused.) In any case, this article is from Man the Builder by Gösta E. Sanström, 1970, a McGraw-Hill publication.

Cheops is the biggie of the pyramids -- that's for sure. It has a square plan with 755-foot sides. Original height was slightly over 481 feet. The area of the base is 13.1 acres. 2.3 million stone blocks, each averaging 2.5 tons in weight, were required to build this baby. Some blocks weighed up to 200 tons. A typical crane used on building construction in downtown Houston might carry one of these blocks. But to lift it 12 stories would be difficult, if not out of the question. However, a hospital in San Antonio, for which CRS served as a consultant, made use of Zachary prefabricated concrete boxes and required a much larger crane to hoist the boxes. So there are bigger cranes. But our technology is 3,500 years later. We're not so damn smart, are we?

By the time the pyramid was underway, the art of quarrying was advanced. But getting the stone from quarry to job site was no easy task since it had to be transported by barge 605 miles down the Nile. According to records, 400,000 men worked on the pyramid (apparently encouraged by overseers with whips in hand). The project took 20 years to complete. No wonder this article began with a quote from a fellow called Herodotus: "Cheops brought the people to utter misery."

The accuracy with which the builders oriented this structure is amazing. They bull's-eyed on true north. The early Egyptian priests knew their sky and were familiar with how the stars behaved. Most Arabs do today.

Time and climate have treated old Cheops gently. Man played hell with it. Since the Middle Ages it has been used as a quarry for building stone. Originally, the Great Pyramid was cased with polished limestone, but this casing has long since been removed and some of it can be found in buildings in Cairo today.

If CRS had existed 4,500 years ago and had been given the assignment to design and fast-track Cheops, we would probably still be negotiating the last few change orders caused by a revision in codes. If not, we'd probably still be involved in a law suit for an injury from one of the stones dropping on a worker's big toe.

400,000 men working on Cheops. What a great opportunity for a hard-hat salesman.

10. Another surprise was the labyrinth of tombs and temples joining the big three, and of course there were the smaller pyramids -- three by the big pyramid and three by the smaller one for the daughters of the Pharaohs.

11. As the morning sky began to lighten, the color of the desert and the stones seemed to change. We looked down on the city and out toward the picture-post card desert.

12. It was a sacrifice to face the world so early this morning, but worth it. We were the only tourists on the site. Great!

13. Nevertheless, out of the sand, it seems, came the hucksters. One with a camel for Aleen to ride. She politely turned him down. After my experience with the damn elevator, I wouldn't dare ride anything more exotic than a taxi.

14. Great morning. But cold. My guess is low 50's. Aleen says 40's. But she's so delicate. A sweater and coat felt good.

15. After we went through tombs and the big temple at the foot of the middle pyramid, "Frank"

wanted his money. He said, "Give me what you think I'm worth." I did. He wanted more. Double.

16. Now about the Sphinx: The original was one piece of stone. Dimensions are gigantic. Not long ago only the head (a king) was sticking above the sand. Now they have excavated the entire bod. The bod is a lion. As we saw it, the Sphinx appeared to be in a huge empty rock quarry. Understand excavation was done in 1925.

17. The Great Sphinx needs a nose job. Napoleon's artillery is credited with knocking it off.

18. I estimated one paw of that lion would just about fill up our bedroom. Quite a pet.

19. Back to the taxi. Goodbye to "Frank" who should have engraved on his pyramid these immortal words, "I tried." Sun just now was on the horizon. What a show ancient man and nature put on for us. We shall always remember this morning. If we had to do it again, we wouldn't change a thing.

20. On the way back in the light of day, traffic became heavy. Right in the middle of a busy

thoroughfare a herd of goats, at least 100 of the creatures, held up traffic.

21. Really cold. Aleen just remarked, "I'm sure glad I brought these warm clothes for England. I would have been in terrible shape for Egypt." She had some long handles.

22. We ate breakfast at the Hilton (Aleen can't take the dining room at the Shepeards for the smell, the dingy uniforms of the busboys and the horrible table linen). On the way to the hotel, we saw a big truck with its too-high load of crates get stuck trying to go through a passageway. The driver was frantically hacking off the top crate to size. Something symbolic about that.

23. Acoustical pollution -- Cairo is full of it. Particularly taxis. Jim Haecker, who finds no wrong in these people, explains the excessive honking this way: "They honk not because they are mad and want you to get out of their way. Not that at all. They honk out of politeness to let you know where they are." That's not the way it sounds to me. They are just aggressive drivers like those in Houston or New York. The

IN **LONDON** THERE WERE NO HORNS BLOWING. NO EMOTIONS? LESS AGGRESSIVE? OR MORE MATURITY?

horn is as much therapeutic as it is a polite warning. More so.

24. After our midmorning breakfast at the Hilton, Aleen and I started a long walk through the thickest business district around Liberty Square. What a dramatic contrast coming from the pyramids directly to a highly-intensified urban space. A 5000 year jump.

25. Another contrast: Walking on the rocks and sand around the pyramids and walking on the elevated pedestrian loop of Liberty Square. Egypt is a country of contrasts -- lush garden areas and deserts; cars and camels; high-rise apartments and half-buried huts; water and sand; centuries-old buildings and modern ones; peace and war!

26. Has man really made improvements? Are we better off today? Is being a slave laborer building the pyramids any worse than being a woman sitting on the curb of a filthy street, nursing her baby, and selling bread? Have we really made progress?

27. Came back to the hotel to rest our flatten-

ed feet before a trip to Old Cairo and the Citadel with some close friends of Jim Haecker. The Shepeard's, by the way, is an excellent building. Our chief complaint is food and the lack of cleanliness in the bathroom. The laundry we sent out came back in great shape.

<u>27 January 75 - Cairo - Late Afternoon</u>

1. Promptly at noon our friends arrived at the hotel. They were Azza Fahmy, the lady silversmith; Nabil Gali, her architect husband; Ahmed Rabie, another architect; and Nehad Nofal of the Egyptian Air Force (flew MIGs). They treated us great; taught us much. Nofal has a sister living in San Francisco. Her husband works for a San Francisco architect.

2. Before we left the hotel, Architect Rabie gave us a most informative lecture on the historical development of Cairo since the 15th century. He's quite a student of Islamic architecture.

3. I want to correct a previous statement. I had said that the ibis was protected by law because it had been the bird of the Pharaohs. The bird is extinct. It's the abou-kirdam, the friend of the farmer, I had in mind. It's protected. OK?

4. This ibis bird must have been precious. In one tomb they found about a million of the birds mummified. The Pharaoh wanted his birds with him.

5. Aleen and I rode in Nabil and Azza's borrowed VW (her brother's). She wrecked their car and it was still in the shop. We drove along a 50-foot 16th century aqueduct which had carried water from the Nile to the Citadel. Saw many leather tanning factories and small shops underneath and behind the aqueduct. Saw donkey carts everywhere. Some were decorated with vividly colored geometic designs. Aleen said she saw similarly decorated carts in Alexandria.

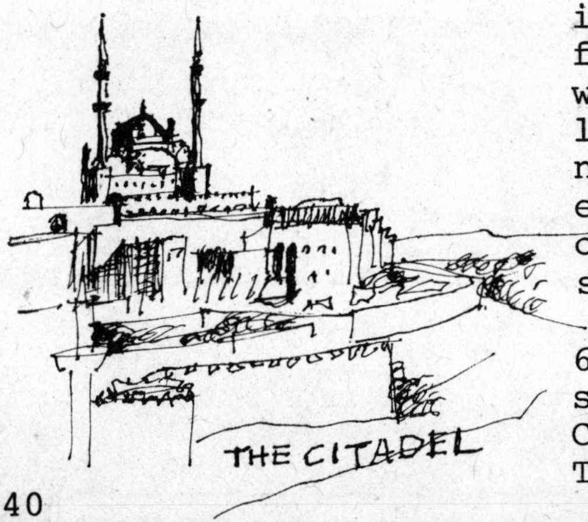

6. All of a sudden the VW brought us to a spot that revealed a most dramatic view of the Citadel high on solid rock. What a fortress. The scale was inhuman it was so big -- for

giants, not for ordinary men. Looked like some of the forts I saw in India. And built about the same period. I bet some architect/ engineer type with a military bent went around the various countries selling plans. The forts were designed to fight off the Christians.

7. Our first stop was to visit the Mosque of Sultan Hassan. Across the street was the Mosque of Refaie, not quite the quality of Hassan's. Ahmed Rabie described the Mosque of Hassan as "the best Islamic piece of all architecture." It's 15th century. Azza Fahmy said it was touted as being "the most beautiful Islamic building in the world." It functioned as a mosque, a school and a tomb. Sultan Hassan was buried there.

8. The mosque had a great court with walls 50 or 60 feet high. Before entering, we put cloth coverings over our shoes as is the custom. Then we went through a dark, oblique hall and when we arrived at the great court, the space nearly exploded. The spatial sequence was most dramatic.

THE **MOSQUES**, WITH THEIR TALL MINARETS, DOMINATE THE SKYLINES OF EGYPTIANS CITIES, TOWNS, AND VILLAGES. THESE MOSQUES ARE BEAUTIFUL, **CLEAR ARCHITECTURAL STATEMENTS**. I LOVE THEM.

THE **MUSHRABIYA** (WOOD LATTICEWORK SCREEN) PERMITS NATURAL LIGHT TO ENTER TOP PORTION AND PROVIDES **VISUAL PRIVACY** AT "LIVING" AREA.

9. Architect Rabie explained the subtleties of Islamic buildings with great enthusiasm. He loves these old buildings and is quite an authority on them. He feels new buildings fail to recognize the technology of natural lighting and natural ventilation and human scale developed throughout the ages. I believe him.

10. Bad news. I was told by another source that the stone of this great building was taken from the facing of the pyramids. I understand some of the churches in Rome were built from the stone "borrowed" from the Colosseum. More cases of robbing Peter to pay Paul.

11. The Mosque of Hassan had beautiful detailing. To eliminate the hard line of building and sky, the top of the building had a lace-like trim going around the edges. Personally, I love all kinds of "sky connections." Silhouettes of buildings, like silhouettes of people, set the character.

12. Religious students lived in the five-story dormitories (not now; about 200 years ago) within the mosque. The stone stairs were pure structural inspiration. The stairs consist of

a system of 14 interlocking arches resting on each other. Most ingenious.

13. Islamic buildings are full of spatial surprises. Frank Lloyd Wright must have been there.

14. Natural lighting and natural ventilation had reached a peak of advancement by the 18th century. We saw evidence of this by visiting two old homes (now used as museums).

15. Natural lighting? Extensive use of top lighting, clerestories and high windows. And very high ceilings. Spaces are a credit to Louis Kahn who seemed to have specialized in natural lighting. So did Le Corbusier, specifically in the Ronchamps Chapel. Frank Lloyd Wright was no slouch in natural lighting either. The Islamic architects, however, had subtleties more advanced. For example, they used marvelously detailed, intricate wood screens which covered window openings. The lower portion of these wood screens was tightly designed to allow maximum visual privacy while the upper half was loosely designed to allow maximum light to enter the

SAME TYPES OF **SUN CONTROLS** FOR HOUSTON CAN BE USED IN CAIRO.

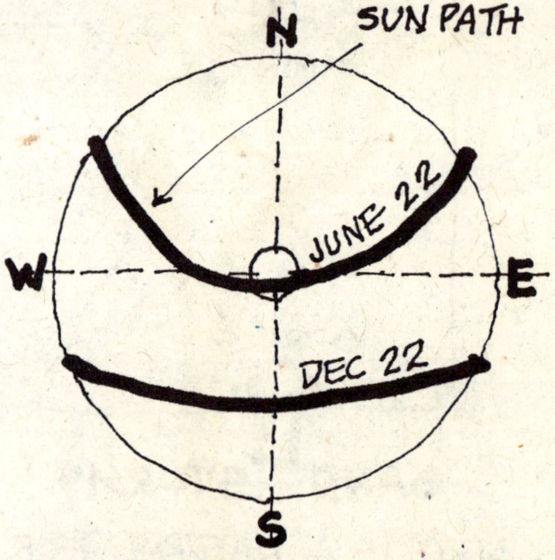

CAIRO'S SUN BEHAVES ABOUT THE SAME AS THE SUN AT HOUSTON. BOTH RELATE TO **30° NORTH LONGITUDE.** IN FACT CAIRO IS SLIGHTLY NORTH OF **HOUSTON.** HARD TO BELIEVE?

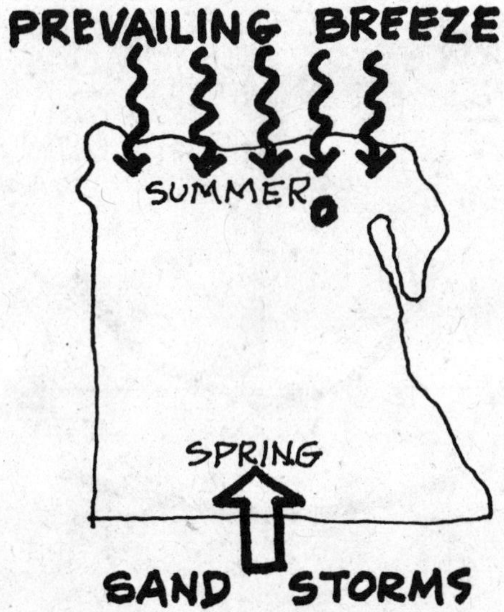

CAIRO IS A "NATURAL" FOR NATURAL VENTILATION. DURING SUMMER, HUMIDITY IN CAIRO CAN BE AS HIGH AS 77 PERCENT.

space, at the same time permitting air into or out of the interior space. These lattice-work screens are called mushrabiyas.

16. Ventilation? I saw my first malkaf (this was in a 17th-century house). Had heard about that ever since my Texas A&M University days when I dedicated most of my life to studying how the sun and wind behave around buildings. In Cairo's Shepeard's Hotel, there were two old engravings which showed an Islamic city with malkafs all over the cityscape. After all these years, I've finally seen one. And what is it? A malkaf is a natural ventilation device which is integrated into the total composition of the building as an important form. It literally says, "Air comes in here." In essence, it's a big scoop on the roof which diverts the wind downward into the interior to cool the occupants.

17. Each house and dormitory in the mosque had ventilation courts. These flue-courts exhaust the air while the malkafs and certain windows and doors serve as inlets. I noticed the modern high-rise apartments had these courts for ventilation. I think they are required by law.

18. Islamic architects were most sensitive to "human scale" for want of another word. Every group of open spaces had a change of floor level with a lower level generally in the center. Spaces were grouped around the lower level as if it were a court. Sitting on the level change was quite feasible. Islamic architects accented spaces rather than rooms because functions were not separated; i.e., eating, living and sleeping were in the same general space.

19. One of the two houses we saw had two courts -- one for living and the other as a kitchen/garden court. Both houses had fountains in the interior living spaces. It's my guess this was done to increase the humidity. Reminds me of some of the ancient palaces of moguls in India in which fountains and running water moved through the house to raise the humidity for comfort. The moderns of all countries, including Egypt and our country, lost the fundamentals of designing buildings using the wind and the sky for human pleasure and comfort. Very sad.

20. The Islamic architects provided high ceilings to facilitate natural lighting and natural ventilation, but they also provided ways to keep

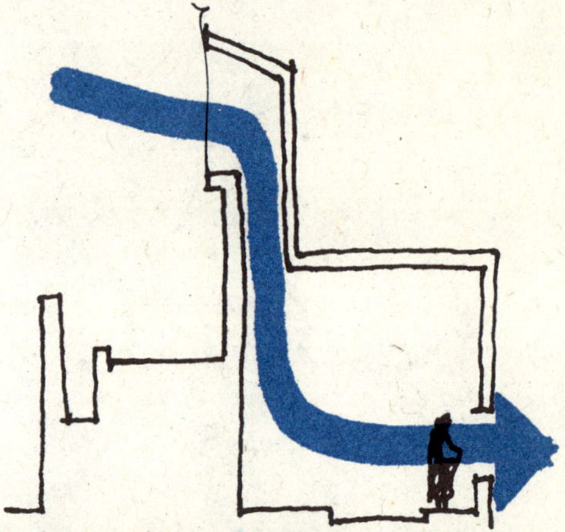

THE **MALKAF** IS A SCOOP TO CATCH THE **WIND**, DIRECTING THE AIR DOWNWARD INTO INNER SPACES. A GREAT VENTILATION **DEVICE** ALLOWING PRECISE DIRECTIONAL EXPOSURE TO PREVAILING COOLING WINDS YET GIVING ARCHITECTS FREEDOM TO ORIENT BUILDINGS FOR OPTIMUM DAYLIGHTING, DESIRED VIEWS AND SUN CONTROL.

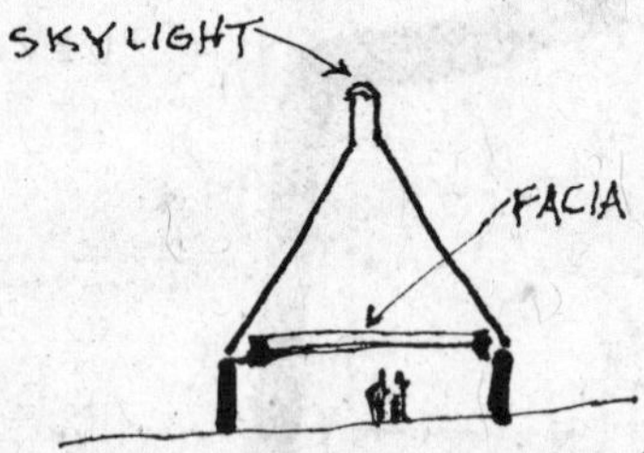

the users of these spaces from feeling they were in a well. One design technique (Frank Lloyd Wright used it. Even Bill Caudill used it in his own house.) was to have an interior facia at "human scale" with the rich details below that line. One is not conscious of the very high ceiling; nevertheless he enjoys the thermal and lighting comfort the high ceiling offers. Islamic architects were most ingenious as well as artistic.

21. Another characteristic of Islamic buildings is their asymmetry. When a mosque is designed and the people praying must face east toward Mecca, then the architects simply twist the mosque to do the job. Symmetry is not precious. The designers had no hesitancy in using oblique corridors if making them at angles increases circulation efficiency. Cutting off a corner of the building to make a pedestrian street more efficient is cricket. Proper orientation of various interior spaces supersedes symmetry. And the Islamic architects had the skill to handle asymmetrical forms. Not many contemporary designers have it.

22. I admire the design skills of Islamic

architects. Any mediocre designer can achieve visual unity by symmetry. Much greater skill is required to work in asymmetry.

23. On our way to and from Old Cairo we saw the city of the dead -- the graveyard. It's unique in that they build what appears to be small buildings and mosques. It does look like a miniature city for dead bods.

24. We visited the market area in Old Cairo and the bazaar (antiques and jewelry). A sight to behold. It's called the kanelkallily. Thousands of craftsmen make everything from plumbing fittings to exquisite gold jewelry.

5. In some areas of Old Cairo, Aleen was shaken by the filth and squalor and "meanness" of human living. To her it was appalling, the worst she had ever seen. Not me. I've been to India.

26. We passed store after store, in fact whole blocks, selling bolts of beautiful cotton fabrics. Aleen said, "My sister, Betty, would lose her mind" over so much and so many varieties of fabrics. Betty teaches sewing.

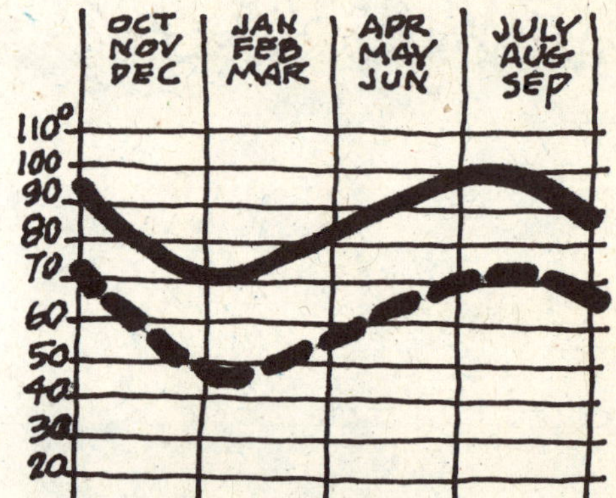

TEMPERATURE IN CAIRO WAS WONDERFUL WHEN WE WERE THERE IN JAN.-- JUST ABOUT WHAT THE CHART READS. SOLID LINE (▬▬) IS **AVG. DAILY MAX.** THE DOTTED LINE (- - -) IS **AVG. DAILY MIN.**

HIGHEST TEMPERATURE RECORDED IN CAIRO WAS **117°F.** LOWEST TEMPERATURE RECORDED WAS **34°F.**

WE SAW ALABASTER VASE-LIKE CONTAINERS THAT WERE USED TO HOLD THE ORGANS OF THE DEAD PHARAOHS OR NOBLES.

27. What a contrast in Old Cairo seeing the beautiful old 18th-century elegant houses with ivory inlaid doors, the best inlaid marble floors and artistically perfect wood screens located right in the middle of poverty -- mud streets, animal excretion, stench.

28. Aleen will be journalist for Luxor in the morning.

28 January 75 - Luxor, Egypt - Aleen

1. We had to get up at 5:00 this morning to catch a plane for Luxor, up the Nile an hour in jet flying time.

2. On the way to the airport, we talked with Rauf, our driver, about housing in Cairo. Rauf, a young man of about 30, remarked, "We have forgotten how to build buildings, but we haven't forgotten how to fight." I commented on the favorable press Sadat had received in the U.S. and Rauf replied, "Yes, but Sadat is getting impatient. The U.S. is the only one who can make

Israel give back our lands."

3. Egypt Air got off an hour late because of a large group of French tourists who really intimidated me. We were the first at the gate and the last to board. Bill says I should be more competitive.

4. While waiting for take-off, I had time to reflect on my visit to the Egyptian Museum while Bill was working. It is a big (three-story), cold, poorly lighted building holding those magnificent artifacts. Few pieces have any kind of identification or explanation. The place looks like a warehouse, but what a wealth of treasures it holds. Almost the only information available was that the columns and various objects came either from the Old Kingdom, 2700-2200 B.C., the Middle Kingdom, 2100-1786 B.C. or the New Kingdom, 1575-1087 B.C.

5. For an extra fee you could see the 20 to 25 mummies in one room; if you had the stomach for it, that is. Didn't want to miss anything, so I decided to chance it. The amazing thing to me was that the linen that bound the bodies was like new. And after some 3,000 years, teeth

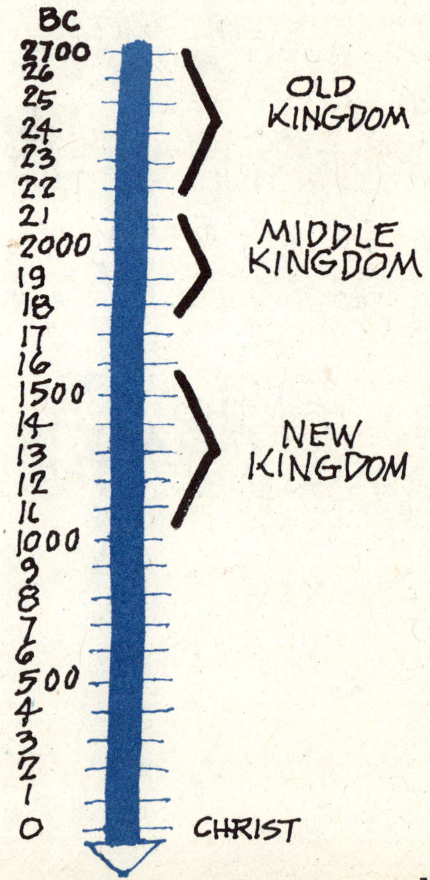

AUTOPSIES ON WELL-PRESERVED **MUMMIES**, SOME 2700 YEARS OLD, REVEALED THAT **HEALTH** PROBLEMS BLAMED ON MODERN INDUSTRIAL SOCIETY SUCH AS ARTERIAL DISEASES AS ARTERIOSCLEROSIS WERE COMMON IN ANCIENT EGYPT.

THIS CLEARLY DEMONSTRATES THE HIGH STATE OF THE **ART OF EMBALMING** OF THE ANCIENT EGYPTIANS.

were still in the mouths and hair still on the heads of the mummies -- some record. Took a close look at five or six and came out with my stomach still okay. 

6. Egypt Air finally got off exactly 3 1/2 hours after we got out of bed. Had our first cup of coffee. Breakfast was a cold omelette, a cold roll and a cold pudding of some sort.

7. From our window seat we could see the undulating sands of the desert -- seemingly endless but as we neared Luxor, we could see a mountain range in the distance and near the Nile, where the land was irrigated, there were trees and lush green fields.

8. We were brought to the Winter Palace Hotel. Our room is in the old part. This old building is very elegant with wide halls, high ceilings and a beautiful staircase at the entrance. Our room overlooks a lovely garden and the front of the hotel faces the Nile and the green fields across the river backed by the stark mountain range.

9. Bill is glad to get a rest from committee

work. He's still reading Architecture for the Poor. It's beautifully written from what I've read. Now Bill, you take it.

28 January 75 - New Gourna, Egypt

1. Aleen and I decided to deviate from the plan and go see for ourselves what Architect Hassan Fathy's village (from Architecture for the Poor) looked like.

2. Our guide, Ahmed El Baron, arranged the trip for us including a boat trip across the Nile and taxi to New Gourna.

3. Tuesday is the market day of each week, so on the road to New Gourna we saw the farmers displaying and selling their wares -- including potatoes, tomatoes, turnip-like things, onions, eggplants, pigeons, sheep and goats.

4. Women all wore black garb. Men wore a variety of clothing.

5. As our driver darted among the donkey carts, honking continuously, Aleen screamed, "Bill, look at the babies." I couldn't see any babies. Then after she pointed them out, I saw what she saw -- two little lambs. I asked her if she was homesick for Ben* and Julie (her son and daughter). ↑ BEAUTIFUL LIKE HER MOTHER

6. Women carried everything from baskets of vegetables to water jugs on their heads. Must be an international mode of transportation.

7. Aleen saw about eight women walking together carrying things on their respective heads. She said, "I wonder what they're talking about. I bet one is saying, 'Here I am wearing my head off down to the bone and my husband is down at the mud cafeteria eating beans and sipping tea.'" He probably was.

8. Yes, we had tea at the mud cafeteria in the center of the marketplace. Our guide, Ahmed, (we think he would much prefer to be with these villagers than waltzing disgruntled tourists around for his uncle, El Baron) invited us to come in and relax over a cup of tea. Aleen acquiesced. I could have killed

* BEN LOOKS LIKE HIS DAD WHO WAS MY ROOMMATE IN COLLEGE.

her! Me with my delicate stomach. Bless her sweet heart, she's so people-conscious she would never decline an invitation. Not me. I'm not one who would never hurt a fly. Nevertheless, we drank the tea -- flies, bugs and all. Lomotil, here I come.

9. EMPATHY HIT ME THE MOMENT I SAW NEW GOURNA. Designed by the internationally famous Hassan Fathy and built nearly six years ago, it's been a controversial project. Now we saw for ourselves. Terrible sight. How disappointed Architect Fathy must be, seeing the place as it stands. Most of the windows in the houses have been bricked in. Some of the arched doorways are squared up with mud bricks (no aesthetic consideration in form and finishes) for rectangular doors. Crude mud fences detracted from the original smooth finished walls. Many buildings are "broken" as one of the young men of the village tried to tell us in pieces of English. The school -- completely destroyed. The "police station" -- abandoned for a better, more conventional white structure. An arcade shop building was not fully used. Houses were distorted with changes.

THE VILLAGE SQUARE WAS SKILLFULLY DESIGNED BY ARCHITECT HASSAN FATHY. IT HAS A INTIMATE HUMANISTIC SCALE.

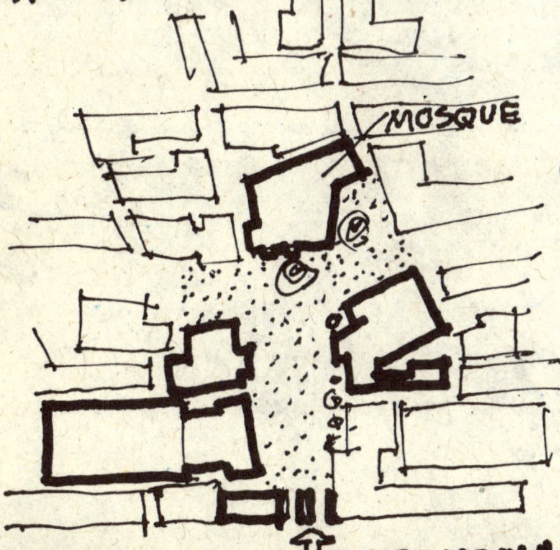

LAYOUT OF NEW GOURNA VILLAGE NEAR LUXOR.

10. From what Aleen and I could glean from the villagers and Ahmed, these things happened:

- The school and other buildings were destroyed by underwater conditions. Simply a bad location for buildings.
- Houses cracked and were removed to give the land back for agriculture.
- The "police" simply wanted some better facilities. Mud was not for them; however, the building was badly damaged.
- Since some of the buildings cracked because of unstable soil, the villagers were scared to live in them.
- The mud brick homes with so many windows were too cold for the people who were used to living in caves on the hills. It's quite possible that economy made having wood or glass windows impossible. In any case, the openings were mudded up.
- The people liked their caves in the hills better because they made their living digging for ancient artifacts. Moving to the valley was a bum idea to them. Who wants to be a farmer?

11. The reason for building New Gourna was the

Aswan High Dam. Since it controlled the floods, the valley could be made habitable. The government moved Old Gourna to New Gourna. Not a totally successful transition, for sure. I suppose no one, including Hassan Fathy, clearly understood the people involved. When they started putting mud on his white walls, it must have hurt him deeply. But to these mostly cave-dwelling people, mud is beautiful. Aesthetic theories must also go along with theories of construction and environmental controls.

12. If one can objectively pull the original village form out from the squallor, destruction, patching and additions, he can see that Hassan Fathy is a first-class designer. The architectural forms were beautifully proportioned. Spaces great. Mud brick technology without using forms very exciting. Outside spaces were nicely done and the conditioned outside spaces such as arcades, porches and courtyards were beautifully executed and intelligently conceived.

13. Some of the buildings, including the mosque, the "police station" and the shops, were expertly designed. It's so sad that things hap-

THE VILLAGE SQUARE AT NEW GOURNA NEAR LUXOR.

pened as they did.

14. I left New Gourna feeling bad. This could have been a great, significant project. Unfortunately both the people and the underwaters of the Nile rejected it.

"THE COURTYARD BECOMES THE OWNER'S PRIVATE PIECE OF SKY"
HASSAN FATHY

15. Before we caught the boat back to Luxor, our guide friend, Ahmed El Baron, ate breakfast in the mud cafeteria. He loves the place and said the food was excellent. And you could tell he loved the people. Aleen called my attention to his position while eating (squatted down, camel fashion). I'm glad I'm not Egyptian. I can't bend that far.

16. While I was pointing out to Aleen a hidden angle iron in the roof of the cafeteria amid the canework, she spotted silver bracelets on the ankles of some of the women (like they wear in India). We have different eyes. If I am more interested in angle irons than ankles, I should have my head examined.

17. Later I checked with another source as to the reason for the failure at New Gourna. This person said the buildings would have worked fine

in Old Gourna; better soil and climatic conditions for mud brick houses. Some people refused to move to New Gourna. Some even had better houses. The reason they blocked up window openings was because they had no wood windows (solid shutters). Said my source, "It was a failure, but some good came of it."

18. That's the best summary yet. NEW GOURNA WAS A FAILURE, BUT SOME GOOD CAME OF IT.

28 January 75 - Luxor

1. Luxor was ancient Thebes. Today we visited the Temple of Luxor and the Temple at Karnak. This time we had a superior guide -- Nagi Samaan, a native of these parts who knows his local temples.

2. To get to Karnak, Aleen and I rode in a delapidated horse-drawn carriage pulled by a dissipated old gray mare. The guide and another tourist from Rhodesia rode in another carriage. The two drivers were most competitive. We felt

we were in a chariot race. The lead changed two or three times. Our contract didn't call for that thrill.

3. We did get a magnificent view of the Nile -- with the green, irrigated lands on the other side and the great rock hills as a backdrop. Very dramatic. We think we spotted one of the temples near the rock cliffs.

LATER WE LEARNED IT WAS TEMPLE OF QUEEN HATSHEPSUT.

4. Arriving at Karnak and seeing the pylons (each side of the gate) at the entrance was just too much. The super scale is hard to imagine. Can you visualize a 13- or 14-story building on either side of the entrance. That's how big those sloping walls at the entrance structure were. Why sloped? Don't know. They looked rather strange with the giant vertical columns inside the wall. The wall itself, confining 300 acres, is made of mud brick and it's 90 feet high, still standing after centuries.

5. What's Karnak? It's a temple of temples, started about 2200 B.C. Continuous work on additions, subtractions and modifications lasted about 2,000 years. Was dedicated to Amon, the God of the Sun.

DIG THIS ONE! 2,000 YEARS UNDER CONSTRUCTION.

6. Try to visualize columns which are over six stories high. The base is as wide as our bedroom and it's a big room. The shaft itself is about 11 or 12 feet in diameter. The guide said that 100 men can stand on top of the capitol. That's a hell of a column. Why the big scale? The temples are for gods, not men. The poor Egyptians never got in the place except for the few who were servants. During the heyday of this temple of temples, there were 8,000 employees, priests and helpers to keep it up.

7. Most temples had their own water supply. It was supposed to be holy water, but the chances are the water was used to irrigate the plants and keep the temples inside the 300 acres spotless. The Shepeard's Hotel in Cairo could take a lesson. This one in Luxor, the Winter Palace, is much cleaner.

8. One of the kings, Ramses II, really made his mark in Karnak. For that matter, all over Egypt. The guy ruled for 67 years. Died when he was 97 which is remarkable since he had 170 children. He must have died from overwork. As the song goes, "Nice work if you can get it."

TODAY--MY MOST VIVID EXPERIENCE IN **PLACE** AND **TIME.** **HERE** I AM IN **KARNAK.** EXPERIENCING **FORM** AND **SPACE** OF THIS GREAT ANCIENT TEMPLE-- IT'S STILL STANDING, ERECT AND STRONG. AND SOME AREAS IN IT NEARLY **4000 YEARS OLD.**

IT WAS HERE--WHEN **MOSES** LED THE **EXODUS** FROM EGYPT. MY BOD IS HERE. BUT MY MIND GOES BACK TO MOSES. THAT'S ABOUT 3400 YEARS BACK IN HISTORY. KARNAK IS **THEN.** BUT KARNAK IS **NOW.** BECAUSE I AM NOW-- EXPERIENCING THIS MAGNIFICENT "TEMPLE OF TEMPLES".

KARNAK TEMPLE
AT LUXOR--HALL OF PILLARS

9. Even King Tut, the relatively unimportant pharaoh who died very young, left his mark on his beautiful small temple with exquisitely carved columns. Some were of lotus blooms. Others were papyrus blooms. King Tut had the artistic touch.

10. During the 2,000 years of construction, Karnak changed considerably. If you guys from Construction Management, Inc. are reading this, they didn't have fast track. 2,000 years to build!

11. Gad, but they built well! And built for beauty. Looking at Karnak makes me wonder again about man's progress. Undoubtedly they had better workmanship, better building skills than we have today. And more lasting materials. We saw some painted bas-relief that has lasted 3,000 years (and in the sun). We saw solid granite obelisks; one piece, mind you, nine stories high. That's a chunk of stone. And beautiful figures that are as modern as Bill Caudill's 20-century taste. And I like modern simplicity.

12. Did you know why the bas-reliefs are always rigidly looking in one direction? The

guide told us that the Egyptians could sculpt in three dimensions and they did when they sculpted the gods. But when they sculpted mere mortals, they had them looking at a god. Sounds logical to me.

13. Where there were columns, and Karnak had thousands of them, there were roofs. You guessed it -- out of great slabs of stone. Some we saw had double roofs with air space between. The lower ceiling stones were ornately carved. Beautiful work. It's hard to imagine how they got those stones up 40 or 50 feet. All carving was done with the stones in place.

SOLDIERS OF QUEEN **HATSHEPSUT** ARE CARVED IN BAS RELIEF IN **PROFILE** ACCORDING TO STRICT CONVENTION.

14. Our guide believes strongly in this theory: Columns were laid up in sections. After the first section was set, the workers brought in fill-earth and mud brick ramps to get ready for the second section. Same procedure for third and fourth sections, gradually filling up the building as the walls and columns were laid in place (rough). By the time the walls and columns had reached their required height, the temple was completely filled with earth, then the stone ceilings were put in place by ramps. Next the earth was removed "out the front door,"

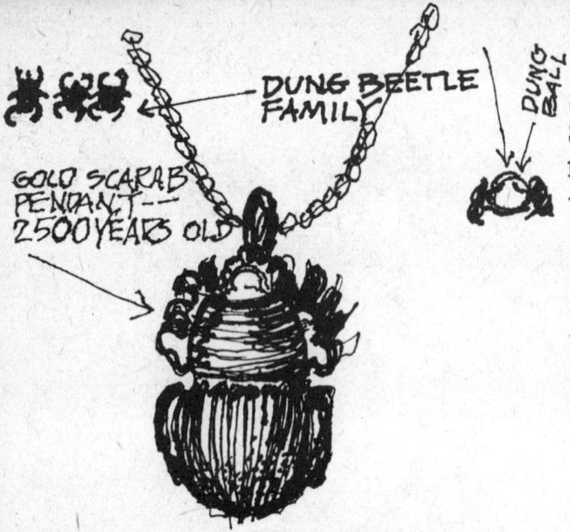

THE SACRED **SCARAB** OF EGYPT WAS A SYMBOL OF **RESURRECTION** AND OF **IMMORTALITY**. AS A KID IN OKLAHOMA I USED TO WATCH **DUNG BEETLES** (SCARABS) ROLL REFUSE INTO BALLS AND PUSH THEM TO THEIR UNDERGROUND HOMES TO BE USED AS FOOD

so to speak. Everything was now in the rough. Scaffolding was then erected and the finish masons and stone artisans were ready to go to work to carve the columns and walls.

15. All the granite, marble and sandstone, with all their carvings, were painted rich colors. What a great sight that must have been. Covering beautiful granite, alabaster and limestone with paint? That tears me up. They are so beautiful without paint.

16. All carvings were done in place as I said before. But think of the number of highly skilled stone carvers needed. Our guide said that boys started this craft (a father to son thing) at six years of age. At 10 they were put on real jobs. At 16 they could be better than their fathers.

17. I said 2,000 years of working on this temple of temples at Karnak involved many kings, each wanting to leave his mark bigger and better than the other kings. Sometimes they would cheat, tear down or cover up the work of other kings. One guy built a wall around one of the obelisks -- to hide the works of his predeces-

sor. That's why it's in great shape today. It remained in its wrapper for thousands of years.

18. The New Kingdom period (Ramses II's time, 19th dynasty -- there were 30) was a boom time. Ramses' day was the peak of power for the Egyptians. Ramses was particularly bad about borrowing stones from others' temples to build his own, or messing up a great court by sticking one of his buildings in it. He literally left his likeness all over the Egyptian world, the big bully. He even made himself a god. I've known a few people like that.

SOME HISTORIAN THINK **RAMSES II** WAS KING DURING MOSES' TIME.

19. Imagine having 2,000 years of accumulated change orders.

20. These people knew architectural refinement. Some of King Tut's small temple columns were pure sculpture. There were two types: those with the lotus motif (round reed) and those with the papyrus motif (pointed reed). These three-to-four-foot diameter columns look as if the reeds were tied together. This notion breaks down the scale. Reminded me of gothic columns. One can "grab" part of the column.

FIGURES OF THE **SCARAB** (BEETLE) WERE MADE IN STONE AND METAL AS SYMBOLS OF **IMMORTALITY.** THE HEART OF A DEAD PERSON WAS REMOVED AND A LARGE CARVED SCARAB WAS PUT IN ITS PLACE — WHEN THE **BODY** WAS EMBALMED.

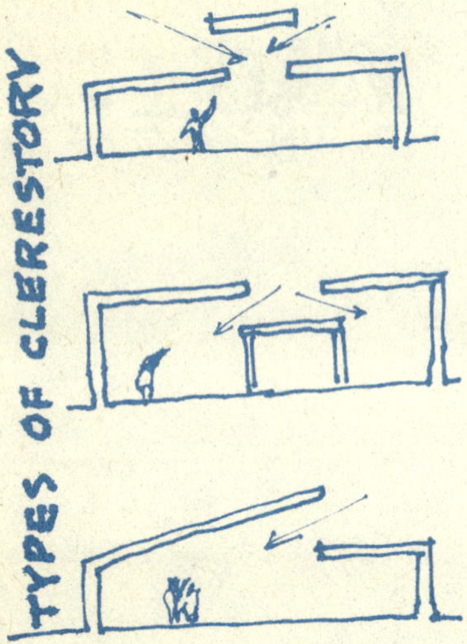

21. Here's something else to think about. In Karnak there is clerestory lighting, designed 1,300 years before Christ. Beautifully done with stone grillwork which lets in high (it is high - six stories up) lighting which filters the sun and bright sky. Those ancient Egyptian architects knew what the hell they were doing -- and they did it with imagination and skill.

22. Aleen and I thought it best tonight that we go easy on the food. Lomotil might strike again, particularly over the senseless tea-partying in the mud cafeteria across the Nile near New Gourna. So we had soup (very good), a cheese dish, and then we skipped the main course and dessert. But Aleen, bless her little stomach, decided she wanted some peanuts. Only where to get peanuts? No newsstand or candy machine. "Ah," she said, "the bar." To the bar we went. We asked the cocktail waiter if we could buy peanuts without the drink. First he said, "No," then after conversing intensely with another waiter, he caught us as we were walking out and said, "Okay." So we waited. Then here he came with a handful of peanuts in a paper cocktail napkin. I gave him 25 piasters (nearly 50 cents, which was too damn much, but

it was the smallest thing I had). The other waiter grabbed it. And the fuss began! Aleen, with her peaceful, compromising disposition and her peanut-hungry stomach, panicked and pulled out an Egyptian pound worth $1.80. I grabbed it and yelled, "NOT THAT MUCH." All eyes were on us -- Aleen, Old Tightwad, Waiter #1 with the 25 piasters and Waiter No. 2 with Nothing. We compromised. Waiter #2 promised to bring back the change. He did. Coins intimidate me. Never was numerical. Can't tell one from another. I grabbed what I thought was 75 piasters so each of the fighting waiters would be equally wealthy, then Aleen and I escaped to our room with our pitiful handful of peanuts. The post mortem of this fiasco revealed that it cost us over $1.50 of hard-earned American dollars for a few lousy peanuts. And we left the waiters still fighting. If Aleen even mentions peanuts again, I'll scream like the jackasses do. And they are loud.

(Note: I let Aleen read this. Asked her if it was slanted too much my way. Her response: "You wrote it. He who writes it gets to bias it." She's a smart one.)

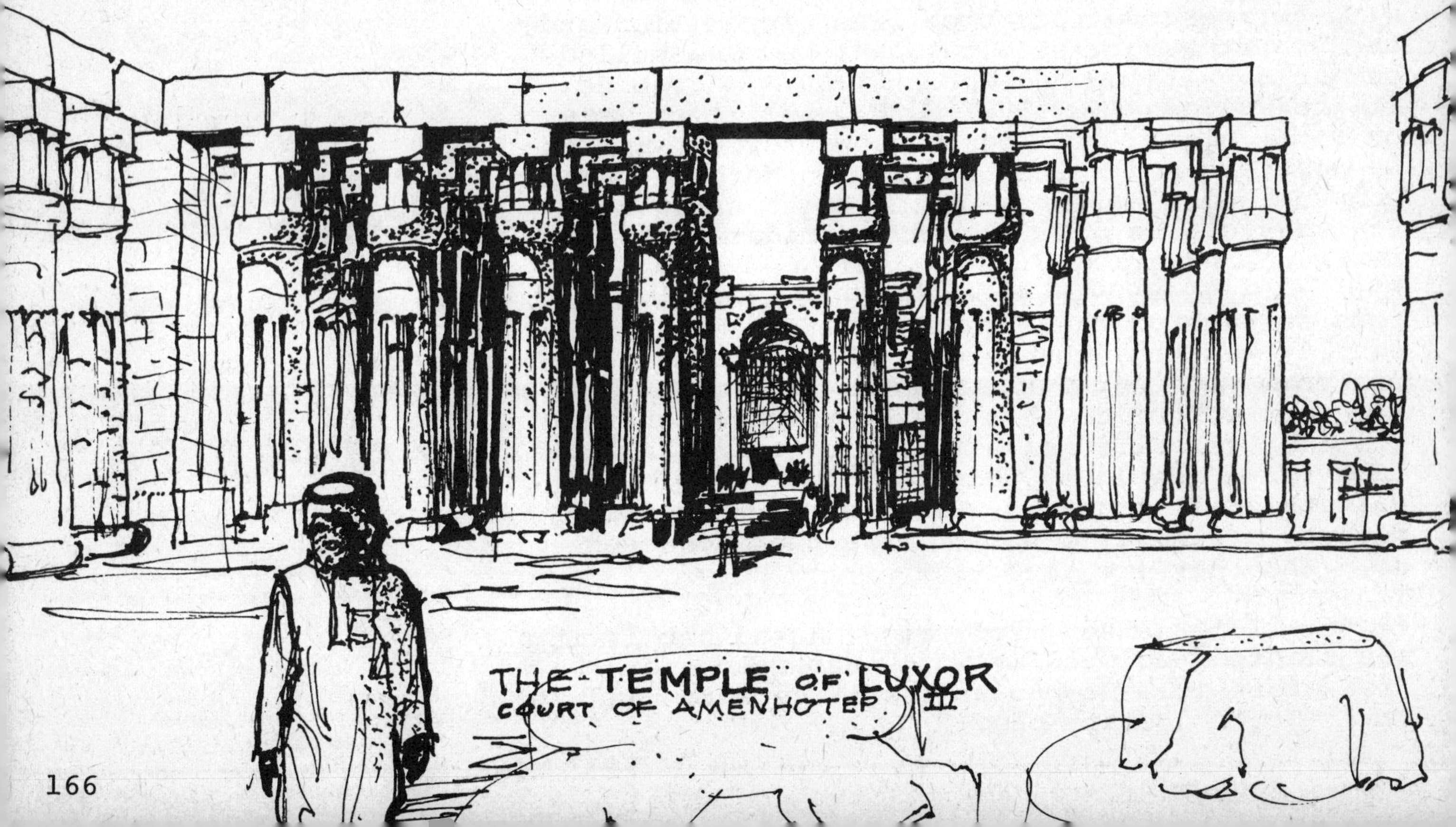
THE TEMPLE OF LUXOR
COURT OF AMENHOTEP III

28 January 75 - Luxor - Before Bed

1. The temple at Luxor is small, 3 acres. It's only 1/100 as large as Karnak (300 acres). Took 200 years to build. Three kings got in on the act. Of course, Ramses was there. That guy gets around.

IT TOOK AS LONG TO BUILD THE TEMPLE OF **LUXOR** AS IT DID TO BUILD OUR OWN COUNTRY.

2. The Luxor temple and those at Karnak were connected with an impressive Avenue of Sphinxes. Our carriage road was about the same route as the original avenue. It seemed at least two miles to me.

3. Luxor was marvelous. Great entrance. Huge sculpture. Time is crazy with sand. A mosque, now over 900 years old, was built on top of a portion of the temple. In fact they didn't dig down to find the temple until a few years ago. Our guide used to live on top of part of the Avenue of the Sphinxes. The mosque is still being used. But the first entrance, which was on sand level, is 40 feet up in the air on one side.

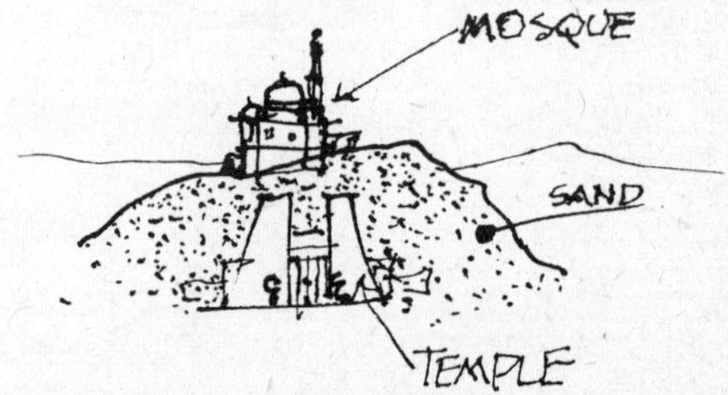

4. The carvings and hieroglyphics are so clear. The Egyptian temples are great history

books. In one wall in the Luxor temple, a front elevation of the temple is carved. Looked complete with flags and statues.

5. Christianity got here through Constantine who tried to make the temple into a church. Christianity didn't last long. The Arabs and Islam took over. The fact that the Luxor temple served three religions is significant.

6. Gad, what a day this was. My feet are killing me. Aleen is holding up better than Old-What's-His-Name. Ramses II is usually portrayed as being three or four times taller than his wife. Put me in stone 1/4 as tall as my wife. But if she ever says, "I want you to meet the little man," I'll erase all this praise.

29 January 75 - Luxor

1. Early morning. Today we crossed the Nile again to its western bank. For Aleen this morning, the sun came up in the West. Some

place between London and Cairo she got spatially disoriented. She concluded that there's a line on the globe where East turns to West and West turns to East. Her logic seems logical. Is West Germany really East and East Germany really West? That's what she said. I think I'm gradually becoming mentally disoriented by her reasoning.

2. Pressed the button for a continental breakfast. No one came. So I stepped out into the hall and yelled to a guy in robed garb down this echo-resounding large corridor. I hollered, "Breakfast." He responded, "Two?" I said, "Yes." Then the hard part: "I want one coffee and one tea." That set the echoes and reverberations bouncing off the walls at 6:30 in the morning. I want my revenge.

3. Why? Last night at 1 a.m. we were awakened by this terrible noise. Pounding. Yelling. Cussing (I think; they did it in Arabic and French). Such carrying on. And doing it at about 95 decibels.

4. Aleen, in her infinite wisdom, decided that someone had gotten stuck in the elevator right

by our door. Good bet. I thought they were waking up someone as was done in London. And as was done to us in Cairo in the middle of the night. In Cairo, after I staggered to the door, two men said, "Is George there? We've got to take him back to the ship." The quiet of the night ain't quiet here, or in Cairo, or in London. And I love that sleep.

5. Today we start our personal tour to the Valley of the Kings, the Hatshepsut temple, the Ramesseum, Madinet, Habu and the Colossi of Mennon. Great way to learn history.

ISLAMIC CREED CONCERNS ONE DIVINE BEING, **ALLAH**, AND THE HOLY BOOK IS THE **KORAN** WHICH ALLAH GAVE TO **MUHAMMAD THE PROPHET** FOR THE BENEFIT AND COUNSEL OF MAN.

6. Daughter Susan, my pretty feminist offspring, are you listening? First the good news: In ancient Egypt the succession to the throne was not through male descendants of the reigning pharaoh, but was determined by marriage to the pharaoh's eldest daughter. Now the bad news: In order to see that a son was made pharaoh, the daughter had to marry her brother. Son Bill, that last part will throw you.

7. Aleen is curling her hair. She said this is a five-curl morning -- just us, the guide and a few mummies. Off we go in a few minutes

across the blue water of the Nile toward the West which Aleen still thinks is the East.

8. One more thing. Our commode or W.C., as they call them, got stopped up. I wish they didn't call them water closets or W.C.'s over here. It's hitting too close. Got a robe-garbed guy to come in to fix it. He came with nothing but a bucket. Used his hand and arm to fix our john. (John Stambaugh, I like that word better.) Believe me, Aleen and I were liberal tippers to this person. He earns his money.

W.C. IS GETTING TOO CLOSE TO MEANING WILLIAM CAUDILL.

9. Flashback to Old Cairo, going through the bazaar. A guy came up, walking fast, darting in and out of the street shops, swinging a lighted brass oil lamp, shouting, "Allah, Allah, Allah." Asked Nabil what this was all about. He said it was an old custom to convey good luck. He added that this is where the Spanish word "ole" came from. "Allah" shouted and "ole" shouted do sound similar. Apparently the Arabs had their influence on Spain in language as well as architecture.

A **MOSLEM** IS ONE WHO PRACTICES THE RELIGION OF **ISLAM**, AS PREACHED BY THE PROPHET **MOHAMMED**. THE TERM "MOSLEM" COMES FROM THE ARABIC WORD **MUSLIM**. NOW ARE YOU STRAIGHT?

10. After breakfast, bright-eyed and bushy-tailed (whatever that means), Aleen and I met

our personal guide, the Johnny-on-the-spot Nagi Samaan, a dignified, informative, real pro. We had him yesterday. Got him through Gaz tours in Cairo. So far Gaz has done okay. Mr. Moose, whom we mentioned before, is assistant manager of the Cairo office.

11. The three of us -- beautiful spouse, Aleen; tourist guide supreme, Nagi Samaan, and the world's most pooped but happy architect, Old-What's-His-Name -- crossed the beautiful Nile again. Aleen said if she had been a pharaoh queen, she would have had a Nile carved into her sarcophagus. Every desert needs a Nile. Every pharaoh should have one.

12. Took a broken-down taxi through the Nile valley to the Valley of the Kings, our first stop. First let me explain my confusion. I had thought a valley was a valley, like the rich valley of the Nile. Not necessarily so. In this case, valley means a carved passage-way in the rock. The Valley of the Kings, a tightly arranged group of tombs for the pharaohs, lies high in the limestone hills. Now you've got smarts too.

13. Taxiing through the Nile Valley, we saw a cemetery for the poor with mud brick markers. Farmers were at work -- loading sugar cane on small, open-bed freight cars, hauling grain and garden products by donkeys and camels. Children were herding goats. Men were riding donkeys to the ferry boat to come to Luxor and parking them in donkey parking lots at the boat dock.

14. Now more trivia, but verrrrry interesting: A camel costs four times as much as a donkey. Camels are more versatile, carry bigger loads, are better at plowing and turning a water wheel. This is Aleen's contribution.

15. Arrived at the Valley of the Kings. There are 62 tombs all built from 1580 B.C. to 1000 B.C. -- covering the New Kingdom -- 18th, 19th and 20th dynasties.

16. Went in King Tut's tomb first. His full name is Pharaoh Tutankhamen. Felt we were walking down into history. I emoted. Kissed Aleen.

17. King Tut came along about 1350 B.C. They made him king when he was eight years old. He

was raised by a priest. Died when he was only 19. Sounds like his whole deal was rigged. X-rays of his mummy indicate that he might have been killed by a blow to the head.

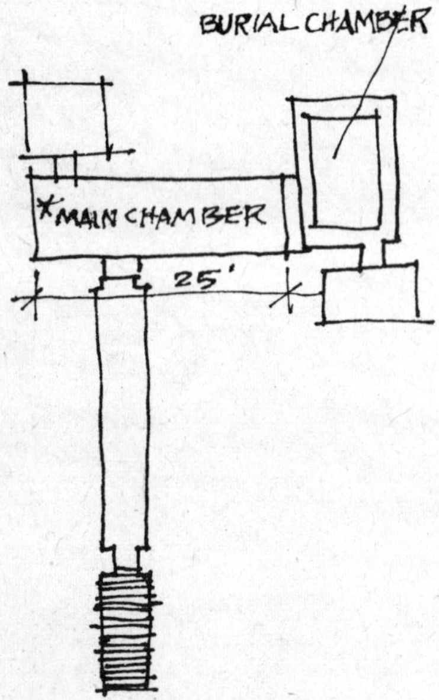

TOMB OF TUTANKHAMEN-- CARVED IN SOLID ROCK, THE SMALLEST OF THE LOT.

18. His tomb is the smallest, but the treasures with his mummy were outstanding. There was one room* in his tomb stacked to the ceiling with goodies, like a warehouse. That was the first chamber Englishman Carter saw when he discovered the tomb in 1922 after six years of digging. Carter spent 10 years cataloging the contents.

19. The significant thing about Tut's tomb is that it is the only one that was untouched by robbers. Just couldn't find it. It was hidden under the tomb of Ramses VI.

20. Here's a parenthetical: The pyramids around Cairo are 1,000 years older than the tombs. The pyramids belong to the Old Kingdom. These tombs belong to the New Kingdom.

21. I've asked Aleen to describe some of the treasures and the mummy cases. There were cases over cases over cases, just like the enveloping dolls I brought back from Russia. Aleen will

write about them tonight.

HERE IT IS.

22. Our guide, Nagi Samaan, believes it was more difficult to build the tombs than the pyramids. It's no small task to carve your way into solid rock, making beautiful bas-relief carvings along the walls of rooms and corridors. Only a few craftsmen could get in a tomb at one time. About 60. 100,000 men, mostly unskilled, could work on a pyramid.

King Tut's body was placed in a solid gold mummy case, the gold case placed in still another one, + then in a third. These cases were put in an elaborately carved stone sarcophagus. Over the sarcophagus was a huge gilded box, over which was a second gilded box, then a third and finally a fourth.

23. There's a theory -- and we saw it work in a tomb without electricity which we visited -- that lighting in the deep interior was obtained with mirrors. Oil lamps would have smoked up and blackened the beautiful bas-reliefs. Lighting deep interiors by concentrated and diffused mirrors beats light bulbs by a kilometer, if not a mile. Wonder if miners in our country ever used mirrors?

24. Next in the Valley of the Kings, we went down into a much larger tomb. King Tut's is tiny -- the smallest. This one is a biggie. It belonged to Ramses VI, 20th dynasty, about 1100 B.C. This one goes down in steps, nearly 1/3 a city block in length. There are all

kinds of chambers other than the one that contains the sarcophagus; then there are the mummy cases; finally the mummy. Centuries ago, robbers got all the goodies which could have required another museum the size of the Egyptian Museum. Workers on the tombs are among the suspects. ☹

25. The Ramses VI tomb had been open for 2,000 years. Yet the beautifully carved murals have retained their color. Some deliberate damage. Don't know whether the robbers did it or the Romans. The Romans and the Greeks, like Kilroy, left their graffitti. The original art work was exquisite. What an artistic culture they had 3,000 years ago.

26. Some of the tombs were unfinished. You could see the sketches on the smooth cut stone walls ready for more carvings. Looked like a Frank Lawyer or a Tiny Lawrence had been here. Beautiful line drawings.

27. Ceilings, too, were painted solidly, but not carved. One skinny goddess, whose stomach was full of stars, ran down the corridor for about 70 feet. That's some stomach.

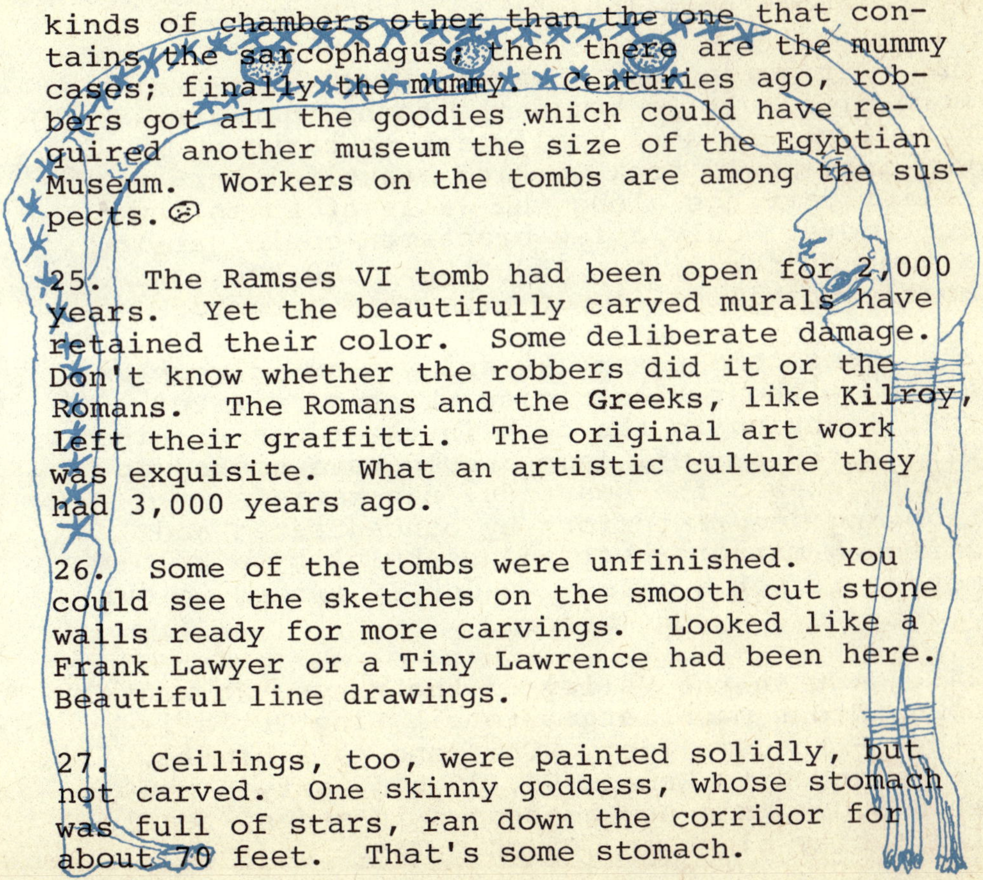

TOMB CEILING PAINTING OF **NUT**, THE SKY-GODDESS CARRYING THE **SUNS** AND **STARS** IN HER BODY THROUGH THEIR COURSE WHEN THEY DISAPPEAR FROM THE SKY.

28. The next tomb was even bigger -- the tomb of Pharaoh Seti, 19th dynasty, about 1300 B.C. Hope I got my number right. So what if I miss it by a few thousand years. (America is only 200 years old.) This tomb is over a city block down, including 110 steps. Seti was the father of Ramses II. You remember that's the guy who put temples all over Egypt.

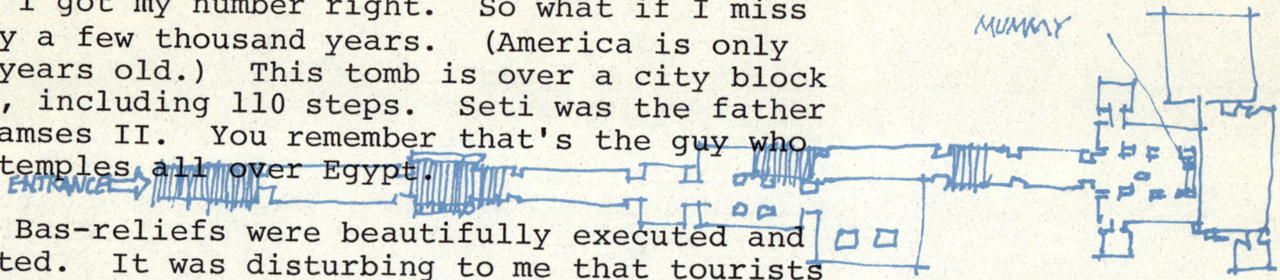

TOMB OF SETI I

29. Bas-reliefs were beautifully executed and painted. It was disturbing to me that tourists could rub their dirty hands on them. No protection. The dirty Christians who used to hide in the tombs didn't help them any either. Soot from their fires caused considerable damage.

30. This tomb is 30 times larger than King Tut's. King Seti had longer to work on his tomb. He lived longer. Apparently each pharaoh had two things to do in his lifetime. Build a temple. And build a tomb. This one was rediscovered in 1817 (100 years before Carter found Tut's) by an Italian. Can't remember his name. But the tomb robbers beat him. He did find a large, beautiful sarcophagus. Sold it to an Englishman. It's now in the British Museum. Saw it a few days ago.

THE **PHARAOHS** BELIEVED IN EVER-LASTING LIFE. THERE WERE IFFIES, HOWEVER. **IF** THEY OBTAINED THE HELP OF THE SUN GOD. **IF** THEY FOLLOWED THE "BOOK OF THE DEAD" TO THE LETTER. **IF** THEY ENTERED DEATH WITH THEIR VALUABLES AND POSSESSIONS (TO USE AFTER DEATH). THIS WAS THEIR ULTIMATE HOPE. THUS THE IMPORTANCE OF THE TOMB.

31. In Seti's tomb there are literally acres of exquisite bas-reliefs. And in vivid colors.

32. Next we taxied to the temple of Queen Hatshepsut, 18th dynasty. She ruled about 1500 B.C. Very impressive. Looks as if it's a part of the huge stone cliff. But it's not dug out of the rock like the tomb. It was built piece by piece. It's now being restored by a Polish expedition. Putting the thing together is like trying to work a giant jig-saw puzzle. In 1885 a Frenchman started its excavation. At one time the temple was completely covered with sand and rock and a monastery had been built on top.

33. A most interesting thing to me was that Queen Hatshepsut's temple had fluted columns similar to the Greek Doric order which came 1,000 years later.

34. Queen Hatshepsut was a peaceful queen. No wars in her time. She carried on extensive trade with Somaliland. The bas-reliefs, which are beautiful, show her sailors bringing back animals, birds and large balled trees.

35. The temples are impressive, but the "man on

QUEEN HATSHEPSUT'S TEMPLE
LUXOR

HERE ARE **QUEEN HATSHEPSUT**'s BOYS CARRYING A BALLED TREE FROM SOMALILAND

the street" never came near the place except to deliver groceries. Only royalty, priests and the clean-up crew ever saw it. I feel privileged.

36. Queen Hatshepsut had her discrimination and equal rights problems, too. Queens were not very popular. When her "likeness" appears on the bas-reliefs, it's in the form of a man. Her successor rubbed her name out and put in his. At least that's the story we got.

37. On the way to the Valley of the Queens, we passed Old Gourna, situated high in the rock hills. Aleen remarked, "I'd rather live there than in New Gourna." Later we saw Hassan Fathy's New Gourna from another direction. From this point we saw seven or eight buildings cracked or fallen down. Siting the village was a disaster. Lack of technological knowhow killed another great project. All the beautiful theories and philosophical verbalizing are in vain if the roof leaks or the buildings crack.

38. The Valley of the Queens is another group of royal tombs carved into the rock for the queens and their children, a little less in

quality than the kings' tombs. (I'm getting sleepy. Will have to cut this short.)

39. Next, to the temple of Ramses III. This one had great open courts with huge colonnades and deeply inset bas-reliefs. Everything is strictly super scale. Beautifully preserved, Even to the painted columns, walls and ceiling. I just couldn't believe this was true. I kept hitting the columns with my fist to see if they were real. In this great temple, I felt like I was in a Cecil B. DeMille papier-maché setting.

40. In one court the Christians had removed one column for an altar. They also had destroyed some of the bas-reliefs which depicted the gods. The Christians were pretty destructive at times. 

41. And last we saw the Colossi of Mennon, two giant sculptured figures that once guarded the gates of a great temple now completely gone. These babies are over five stories high. 

42. On the way back across the Nile, our boat was filled with Russian tourists. Last year

DURING A 3-WEEK STAY IN THE SOVIET UNION, I FOUND THE OPPOSITE TRUE. TO ME THE **RUSSIANS** WERE WARM AND FRIENDLY.

some 14,000 Soviets visited Egypt, most of them in large groups taking week-long swings through Aswan and Luxor. A young English photographer on the boat with us remarked, "Most unattractive race in the world." Aleen said afterward, "Anyone of those women can lick me." Aleen will take it from here.

29 January 75 - Evening in Luxor - Aleen

1. No wonder the wealthy British liked to spend the winter here. Looking from our hotel window, I could believe we were in Palm Springs, California -- blue skies, mountains in the distance as backdrop for the palms and green of the valley.

2. The lifestyle here is very different, however. The men and women are wearing kalabias (the long, loose robes) and amaints (head scarves), and the main street of Luxor has more traffic in donkeys than in cars.

3. Saddened by the filth and poverty every-

where, but at the same time irritated by the peddlers who hound you. When we visited one of the tombs, a peddler of souvenirs followed us for a block, then ran alongside the taxi and thrust his arm inside the open taxi window demanding that we buy his little ceramic mummy for whatever we wanted to pay. He had started out asking £5 or about $8.50. He ended up asking for about $1. I had weakened in Cairo and bought a souvenir outside the Cairo museum so I'm sure the Egyptian grapevine had relayed the word that I was an easy mark for the hucksters. I'm sure they'll be waiting for me in Aswan too.

4. Wondered what kind of system produced these tombs with all their artistry and skill. Learned they lived under a strict system of administration and were classified according to their work. The hierarchy was something like this: a. Designers and scribes; b. Artists, painters, draftsmen; c. Quarrymen and masons; d. Mortar mixers and diggers.

5. A strict record of attendance was kept. The written excuse for our school children is nothing new. The workers who had been absent had to present a written excuse such as "had to visit

my mother-in-law," and, of course, illness was a frequent excuse according to historians.

6. The workers were paid each month with charcoal, fish, dried meat, cloth and materials for their work. They lived in small, uniform houses that were sparsely furnished. In their spare time, they made objects and furniture for their own tombs. There was no question in the Egyptian mind about an afterlife and they wanted some of the good things from this life waiting for them when they returned.

7. Egypt Air was late again. I visited the water closet and was ushered in by the male attendant.  This always upsets me even though it's happened several times before. I've learned to be smart about checking for toilet paper and noticed that there was none. I told the man and he obliged me by going to a metal cabinet where he kept the precious stuff, removed a roll and tore off a very small piece -- too small -- which he gave me to take into the closed (thank goodness!) inner sanctum.

30 January 75 - Luxor and Aswan, Egypt

1. Today is the day we go to Aswan. A little apprehensive. Have no confirmed plane reservations from Aswan to Cairo. Gaz tours said, "Okay; nothing to worry about." I worry.

I'M BACK.

2. Should have brought along an alarm clock. I know better, but didn't follow my check list. Room telephone doesn't work, so I left a 6 a.m. call at the desk. My biological clock, with its sensitive anticipation, went off at 3, 4 and 5 a.m. Got up at 5:30. But I shouldn't have worried. Robe-garbed guy brought in our continental breakfast at 5:50.

3. Sitting here watching the sun come up. Beautiful sky. Beautiful palm trees silhoutted against the constantly changing sky. Donkeys braying -- but loud.

4. Some trivia: Our water closet, which doesn't work very well, has a chrome-plated, flexible tube with a chrome nozzle hanging neatly on the wall beside it. Even well-traveled Aleen had never seen one before. It's my guess it's to use in case of the paper shortage mentioned

by Aleen.

5. And speaking of plumbing, our Winter Palace Hotel room has a bidet. Never use them myself.

6. The importance of Egypt's Nile cannot be over-emphasized. Nothing could have happened here 6,000 years ago if it were not for the Nile. We crossed it a dozen or so times here and in Cairo and the more I see of it, the more I love it. Water and desert -- you can't beat that combination. The potential of a Palm Springs. But it runs up hill on the map.

7. My duck -- a Lake Amphibian, Zero Six Luv -- would love the Nile. The water has been so smooth and blue. It's like a constant, level lake. Lake Aviation should get a salesman out here.

8. My brain is still vibrating from what I've seen -- the great temples and tombs. What an advanced civilization the old Egyptians must have had. Stone technology and craftsmanship, the art of sculpturing and painting, construction methods, technology of natural ventilation and natural lighting -- all amazing. I know

it's not fair to contemporary Egypt to do this, but how can I not compare what was done 3,500 years ago with what they are doing today.

PRESIDENT ANWAR **SADAT** HAS LIBERALIZED THE GOVERNMENT OF EGYPT -- A REAL POLITICAL AND ECONOMICAL OVERHAUL OF EGYPT.

9. The highly sophisticated art of stone masonry is lost. So is the art of natural ventilation and natural lighting. Pride in the physical environment seems to be lost, but I don't really think so. The contemporary Egyptians are proud of their speed in getting buildings up and rightly so; Suez' mammoth housing projects are taking less than a year to put up. The temple at Luxor took 200 years.

10. WHAT A CHALLENGE FOR OUR EGYPTIAN FRIENDS TODAY. THEY CAN'T COPY THE PAST. WHO WANTS TO? TODAY IS THE DAY OF THE COMMON MAN. TODAY WE DESIGN AND BUILD FOR THE PEOPLE, NOT JUST A FEW PRIVILEGED PHARAOHS AND THE RICH.

11. THE CHALLENGE, HOWEVER, IS TO BUILD NEW EGYPT ON THE INIMITABLE ARCHITECTURAL HERITAGE OF THE PAST. LEARN FROM THE OLD KINGDOM; LEARN FROM THE NEW KINGDOM; LEARN FROM THE WONDERFUL ISLAMIC BUILDINGS, CHURCHES AND HOMES...THEN CREATE FRESH AND NEW PUBLIC BUILDINGS, HOUSES, SCHOOLS, HOSPITALS FOR EVERYDAY PEOPLE.

THE **CHALLENGE** AS I SEE IT.

ABOUT SADAT AN INSERT

President Sadat has definite plans and dreams for his country, according to Time correspondents Wilton Wynn and Karsten Prager. The following thoughts were expressed by Sadat during an interview last spring:

"Nasser was the first true Egyptian to rule this country in more than 2,000 years. But in spite of all that foreign domination, our personality never dissolved. One of the principal elements in the success of our battle last October was the faith of every Egyptian, the pride he feels for his land, and the very name of Egypt itself."

"If Egypt is powerful, the Arabs will be powerful. We are proud of our land, and maybe some critics see this as Egypt first. But I feel that I cannot make myself understood in the world today without using methods that people elsewhere understand. We Arabs are very hot. We flare up and we cool down. But here in Egypt we now are using language that can be understood all over the world. A man must be a man of his word. I say what I mean, and I mean what I say, and this is not based on sentimental factors or flaring emotionalism but on wise calculation."

"I think our population can be an asset if it is used, trained and educated in the proper way. We must enter the era of technology. We cannot continue being backward, or we will perish like the Indians in the U.S. If the Indians had mastered technology, they would have fought you until this century."

"We shall retain the public sector. It was our backbone during the very dark years of humiliation and defeat (after 1967). It will continue to exist because it can take responsibility for building what the private sector fears to build: industry that may be unprofitable but that must be built. But let the private sector also be active. It will not be an open capitalist system, and I don't share the fears of the extremists. We will welcome Arab and foreign capital to help us build our country."

First in priority is Egypt's rebuilding program: to restore its shattered cities and give the people some early measure of relief. But Sadat has even more ambitious projects for revitalizing the country. In essence, he wants to free Egyptians from their centuries-old dependence on a narrow (average width is 7 miles), 500-mile-long green belt on each side of the Nile River Valley. The completion of the Aswan High Dam, with its water control and vastly increased production of electricity, will be the first step in making that freedom possible.

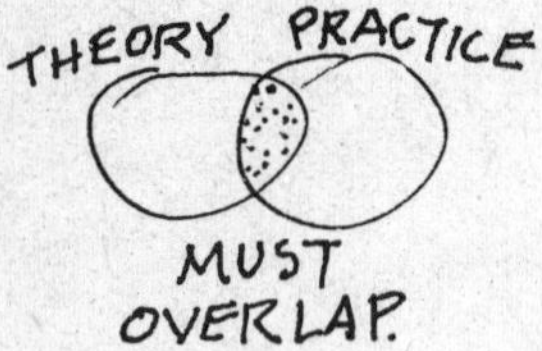

12. Architect Hassan Fathy must have had this in mind. New and needed theory was there. Dedication to an idea was there. Unfortunately the location for New Gourna was a bomb, literally -- systematically destroying the new village like a slow-motion earthquake. Very sad, because the humanistic approach to architectural design was given a severe setback.

13. Looking at the beautiful photographs of New Gourna when it was first built and seeing the village at it is, was as disheartening as seeing the shelled Suez. But it's better for the underground water to destroy the land than man's cannons destroying the people's homes. In both cases it's a disaster.

14. This whole day was one of frustration and suspense. We were taken to the airport in good shape by a Gaz guy, Ahmed Kamil Kamil, who really works for Mediterranean Tours and Travel in Luxor. He's the nephew of his boss, El Baron. He tries. We got the impression he thought it was more fun living in the village, loafing around the mud cafeteria. But he got us back to Luxor's airport and on the plane to Aswan, a half-hour flight.

15. When we approached Aswan, we had a good view of the High Dam which the Russians built, and the earlier dam downstream which the British built years ago. Water everywhere in this desert land.

16. Our apprehension was right. No one was at the airport to meet us. ☹ We asked every person who looked like a travel agent if he was a Gaz guy. And every person looked like a Gaz guy. The little air terminal was crowded with people. Finally Aleen, who can be aggressive at times, pinned down one of the travel agent-looking people who happened to know the man we were looking for. "He's downtown at the air terminal waiting for you."

17. We were never told there were two air terminals -- one at the airport and one downtown on the Nile.

18. So we got on a big bus. "Sorry," the driver said, "this is a private tour." "Where then?" we asked. We finally ended up with baggage in the back of a second bus which was carrying the overflow of the private tour which had rejected us. At every turn the baggage a-

THIS ONE SOLID BLOCK OF POLISHED RED GRANITE CAME FROM **ASWAN**. IT'S THE QUEEN HATSHEPSUT OBELISK, KARNAK.

ACCORDING TO RECORDS ONLY 7 MONTHS WERE REQUIRED FROM QUARRYING, TRANSPORTING DOWN THE **NILE**, POLISHING, INSCRIBING, AND PLACING IN POSITION.

THAT'S FAST WORK! WE COULDN'T BEAT IT TODAY.

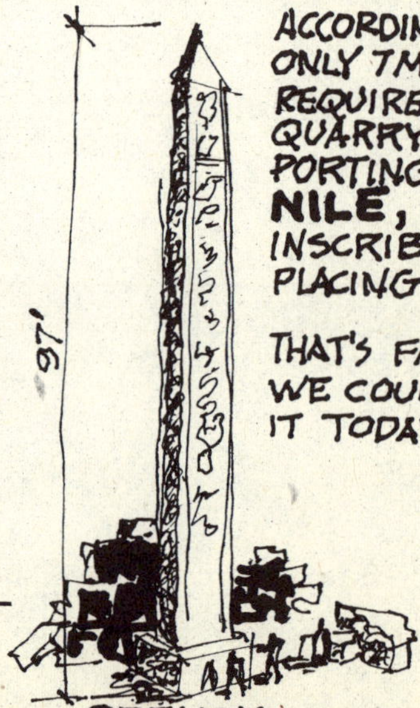

OBELISK

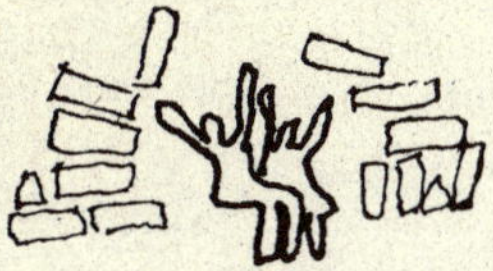

bove our heads kept falling down on us. And there were scores of turns in rocky hills.

19. Aswan, as I've said, is where the granite for the temples and tombs came from. They were floated down the Nile (up on the map) on rafts of huge logs from Lebanon. I suppose the old Egyptians "sailed" the logs up the river or man-powered them.

20. Enroute to the village by the dam, we crossed the old stone English dam with its cataracts to our left and caught a glimpse of the recently constructed High Dam.

21. The bus stopped at some place that might have been a terminal. We weren't sure. As it started up again, I yelled, "Stop the bus." I had just seen a sign that said "Egypt Air." Aleen and I literally fought our way through the baggage into the aisle and out the door. We were the only ones off.

22. Thank God it was the terminal. And there was our Gaz guy. So casual. So unconcerned. Everyone should know Aswan has two terminals.

23. Aleen and I immediately asked him to confirm our tickets back to Cairo later on in the day. We wanted no slip-ups. He sauntered up to the ticket counter with tickets; talked to someone for a few minutes; then shrugged his shoulders and said:

24. "SORRY, ALL FLIGHTS ARE FULL!" 

25. I bit my bullet and calmly said, "But we were promised by Cairo Gaz that space back to Cairo would be certain." "Sorry," Dummy (as Aleen referred to him) replied, "more tourists than were expected." Then he said, "You talk to the man. Tell him you have a connection."

26. I wanted to say my connection was President Sadat, but I restrained myself. He confirmed, not our flight, but our fears: Later flights all filled.

27. The man said he could get us on a flight that left in two and a half hours. Aleen and I said, "We'll take it." Even at the sacrifice of most of the Aswan tour. Gad, we wanted to get back to our "home away from home," the Shepeard's Hotel, even with its rings around the

EGYPT'S MONUMENTAL **ASWAN HIGH DAM** IS A MIXED BLESSING. IN THE LAST 15 YEARS ITS LIFE-SUSTAINING WATERS HAVE DOUBLED FOOD PRODUCTION, BUT THEY HAVE ALSO CAUSED SERIOUS ECOLOGICAL PROBLEMS, NOW BEING STUDIED WITH U.S. HELP. BY ALTERING THE AGE-OLD FLOOD CYCLE OF THE NILE, THE DAM HAS ALSO CHANGED THE LIVES OF THOUSANDS OF PEASANTS.

THE CHRISTIAN SCIENCE MONITOR 7 MAR 75

The Pharoahs were the only persons who were allowed to wear false beards. (See middle figure, opposite page.)

bathtub. To hell with seeing the High Dam, the granite quarries, the unfinished obelisk, and some more temples and tombs -- we wanted to get where we had more security.

28. So we left it to our Gaz guide, Mr. Casual (I didn't inquire what his name was), to see what we could see. He hailed a cab and took us to the Old and New Cataract Hotel where his office was. Said we would take a boat ride to Aga Kahn and then on to the botanical gardens. We would eat a box lunch enroute. And we were hungry. When he tried to charge us nearly 10 bucks for two lousy box lunches, I politely said, " "Stick 'em up your ice box."

29. Our guide took us down to the dock, gave us another guide's 14-year-old son who didn't speak a word of English, and put us on one of those great little sailboats all to ourselves with a two-man crew -- the skipper, a big black man about 28 years old, and his helper, a black boy about nine.

30. Aswan is really another country, occupied mostly by Nubians (blacks). The skipper was a big, strong, gentle man who spoke softly to the

child sailor. Apparently he earned his living sailing this water taxi on the Nile, hauling tourists to Aga Kahn's memorial and the botanical gardens which are on an island. He could say a few words in English. We liked the black skipper.

31. The Gaz guy had said we had plenty of time to see both Aga Kahn's memorial and the gardens. Baloney! Had we paid any attention to him, we would have missed our plane and I wouldn't be in Cairo writing this.

32. Aleen and I decided to go to the island garden without paying our respects to Aga Kahn. Not enough time. Also, the shrine was situated on top of an enormously large hill. My pigeon legs were already overworked.

33. The boat ride on the upper Nile just below the cataracts and the dams was worth the troubled trip. There were hundreds of the happy little sailboats with happy black sailors. And we saw a lot of school children singing and yelling in unison. The sails were beautiful. Water was clear and cold. Great boulders of granite rose out of it.

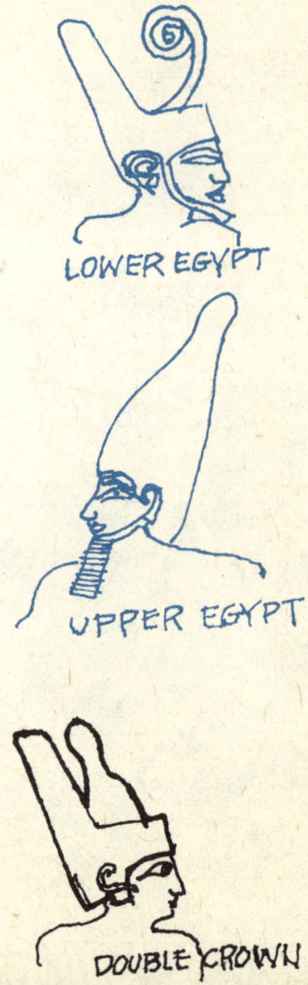

THE **DOUBLE CROWN** SIGNIFIES THE **UNION** OF UPPER EGYPT AND LOWER EGYPT. THE **UNION** OF THE COUNTRY TOOK PLACE ABOUT 3,000 B.C.

NEW CATARACTS HOTEL & NILE

34. Seeing Aswan -- mountains of granite -- made it easy to understand how and why the early Egyptians would go there to get granite for their temples and tombs.

35. The island gardens were not up to Aleen's green thumb expectations. To stave off hunger on the island, we bought a bag of peanuts for the ride back. Got gypped again. During the language eruptions, we both got mental vertigo and paid the little guy about a buck too much. But oh! those peanuts tasted good after my rejection of the $10 box lunches.

36. Sailing back to the hotel was delightful but cold. Scenery unequaled. But we were fighting time. "Wind, push," I kept saying. Pointed to my watch and then to our some-English-speaking skipper. Almost immediately we hit a calm. Of all things to happen.

37. But our big, friendly skipper was unperturbed. Calmly he rearranged a few boards and put his back to the oars. A giant of a man. And we made it.

38. It's a good thing we scratched Aga Kahn

because we just did have time to grab a cab for the downtown terminal. In fact, the Gaz guy had a cab waiting with our baggage already in it. He must have set my new Hartman leather attache case on a board with nails in it because it was all chewed up. Snafu. Snafu.

39. Our Gaz guy taxied us to the downtown air terminal, but that was it. No helping hand whatsoever when we got there. Didn't even tell us we had to go through security. He hardly waved goodbye to us when we drove off in the bus. He won the unpopularity contest.

40. Rode to the airport through the winding, up-and-down streets of Aswan, a favorite recreation spot of the rich British 50 years ago. Recrossed their stone dam, saw High Dam to our left, passed "cities of the dead" (cemeteries) with miniature mosques, saw plenty of military establishments and their tanks and sand-covered barracks, and camouflage all over the place. Also saw a big billboard of King Faisal and President Sadat. Two weeks ago they met here.

41. Got our boarding pass at the airport. Great relief. But we waited and waited. No

plane came in. We began to get uneasy.

42. While waiting, we got acquainted with a nice young couple from Rhodesia, Brain and Moyra Kileff. His brother, Clive, received a Ph.D from Rice University about the time I was there as Director, School of Architecture.

43. Finally a plane arrived -- a large four-engine jet prop Polish airliner. Wasn't sure it was ours -- language difficulties. It was.

44. Again, as so many times before, we were first at the gate and last on the plane. Aleen is so gentle, kind and considerate. I must teach her to push and shove like all the rest of the world's people, especially the French.

45. But back at the downtown terminal I did see Aleen really move. She left me to go to the little girls' room. She'd hardly gone in when she came flying out -- her face beyond the blushing stage. "Bill," she cried, "I've never seen anything like that in my life. The seat was covered with it. The floors were covered with it, even the walls were covered with it." I didn't ask what "it" was. ☺

46. Getting on the Polish airliner was pure, unadulterated confusion. The French tourists jumped the gun so we had to wait at the foot of the airplane steps for another 30 minutes. The attendants couldn't get anyone to go back to the terminal (about a city block away).

47. Aleen felt a bit uneasy about flying in this Polish bird since three of them went down during the pilgrimmage to Mecca. I was ignorant of this grim fact. I did read the sign on the seat in front of ours:

KAMIZELA RATUNKOWA JEST POD FOTELEM.

Translated it means, "Your life jacket is under your seat." To fly down the Nile?

48. Getting back to Cairo was a relief. What a hectic day! But it didn't end. Our taxi cab driver set an all-time record in "broken field running" -- dodging, darting in and out, running lights, gesturing, laughing, and shouting and bragging "Goooood" when we narrowly escaped being crushed by a bus on one side and a truck on the other. One time the driver came so close to a car that he hit its side with his fist.

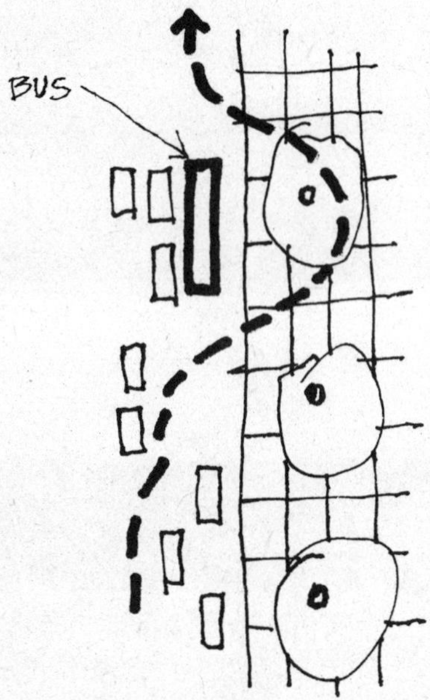

Another time, to get ahead of a bus, he drove on the sidewalk between trees. Aleen got a bit upset with me, saying I encouraged him to do this -- going hog-wild down the streets of a seven-million-plus city. She exaggerated. All I did was to pat him on the back, applaud ever so often and shout "Allah" every once in a while. Allah was with us. We made it to the Shepeard's. Good to be back to Room 818 with its view of the Nile and the accumulated rings in the bathtub.

49. But the end of this hectic day still hadn't come. There was a message for me to call the American Embassy. Scared the hell out of me. Someone back HOME sick or hurt? My imagination ran wild. Called the Embassy. No one there until 8:00 in the morning. What a day!

31 January 75 - Cairo

1. At 8 a.m. I was waiting in the American Embassy to see what the bad news was. What a pessimist! ☹ A beautiful person, Suzanne Raphael,

saw me. No bad news, she told me. Just that they were puzzled. It seems two envelopes with money in them and two names on them -- Neils Thompson and Bill Caudill -- appeared at the Embassy door early Sunday. What for? What to do with the money? Suzanne said, "If you are so rich as to give money away, I can use some." She's cute. I assured her that we were not giving away moola, just returning the stuff to the Embassy -- it was money we hadn't needed and we had been instructed to leave it at the front office. Never did know how to handle money properly. That's why I'm the Chairman of the Board of CRS instead of the President.

2. Aleen and I spent part of the morning with Architect Nabil Gali and his wife, Azza Fahmy, the silversmith. What a beautiful couple -- old friends of Jim Haecker's and now ours. Jim talks of Azza as not only being Egypt's most talented silversmith but as having the most infectious smile and enthusiasm. He's so right. Nabil, too, is a wonderful, quite knowledgeable, fun person. Egypt needs to send these two to America as artistic ambassadors, lecturers and showers of jewelry. They brought Aleen and me an embroidered wall hanging made by children

LIBERATED **MALES** TAKE NOTE: THE BEAUTIFUL **JEWELRY** OF ANCIENT EGYPTIANS WERE WORN BY MEN.

of the village of artists which we visited last week. It's a beautiful piece and we shall always treasure this thoughtful gift.

3. This afternoon, Dr. Hussein A. Saleh of our joint task force picked us up at the hotel and drove us to his apartment for a late lunch. More accurately an early dinner. The food we had! The dinner was Egyptian, plus real Italian lasagna and Maxwell House coffee. What a cook Hussein's wife, Samia, is.

4. These two people lived in the U.S. for seven years. They have a son born in California who can become a U.S. citizen at 21 if he chooses; another son born in Pennsylvania who has the same choice; and a third son who is a local -- no choice. Hussein and Samia have known each other since they were children. In fact they are cousins.

5. Hussein and Samia live on the fourth floor of a nice apartment high-rise on the very perimeter of Cairo. From their balcony they get an excellent view of the pyramids. Looking down on the street, one gets a weird blend of rural (water buffalo, camels, donkeys) and

urban (new VW's driving up to modern high-rise buildings). Their apartment is Americanized.

6. These people are great. They love Americans and America. He's sharp as hell, having earned a Ph.D from University of Pennsylvania. He worked at Vincent Kling's office. Very special. Their kids are beautiful and well-behaved children.

7. At dusk from their balcony we watched the sun set on the pyramids and the farmers bring in their buffalo cows from the field. Great experience.

8. Great day!

1 February 75 - Enroute from Cairo to London

1. This Saturday morning things began to pop early. In fact very early. At 12:07 a.m., to be exact, the phone rang. And who was on the other end. I shouldn't have been surprised. It was Betty Lawyer.

2. Betty had just arrived from Beirut with a group from the American Women's Association on a two-week tour of Egypt. So glad to hear her voice. How she was able to locate us would have surprised me if she weren't enterprising Betty Lawyer. Unfortunately we couldn't get together because Aleen and I had a 5:45 call to make the plane to London. Aleen said, "How nice it would have been to meet Betty in Cairo; as you know, Bill, I've met Frank but not her."

3. Betty said husband Frank and his group in the CRS Beirut office were really working to get out a package for an office building in Dhahran, Saudi Arabia. I was delighted to hear for the first time that we had this feasibility study. I believe Betty said it was for Aramco. My little head doesn't function after midnight.

4. Morning came early. Next time we'll bring our alarm clock. I woke up before call time as usual, anticipating going to the airport and a bit apprehensive as to whether we would get our call.

5. We did. The phone rang exactly the same time as someone knocked on the door. That's

when the pandemonium began. Did you know it took nine people to get Aleen and me (we only had two pieces of luggage) from our room to the special minibus which the Ministry of Housing and Reconstruction had waiting for us. Everyone wanted tips and substantial ones, too. Among those having their eager hands out were porters, maids, houseboys, waiters, elevator boys, floor managers and each one had his own assistant. Some even asked for tips for the people who were not there. ☹

6. During our two-week stay, we had tipped most of these people when we were served. Aleen said not only had they never washed out the tub, but she believes they never swept out the room during our stay. The Shepeard's is a great-quality building, but it's poorly run. And dirty, despite servants all over the place. If it's government-run, and I think it is, nationalization of hotels is a bum idea.

7. Nevertheless, when we did leave, each servant expected a large tip and they were waiting just outside our door. I had anticipated this and had plenty (I thought) tip money. They didn't think it was plenty. After I had ex-

hausted my supply on what I thought was the total group, another wave descended upon us. I panicked, started giving out dollar bills and pound notes, but I couldn't remember to whom I gave what. I'm sure the rich got richer and the poor got poorer. I should have hired aggressive Nabil Ahmed, the man assigned to us from the Ministry, to head up the negotiations for the payoffs. He's rough on the locals.

8. Nabil had the bus at the hotel as usual, 10 minutes early. He epitomizes efficiency. A man of action like his boss, Minister Osman. Happiness is having Nabil Ahmed around to take us through the pushing crowds, around ticket counters, through passport gates, customs, security, etc., etc., etc. ☺

9. Nabil patiently sat with us in the airport until our flight No. 777 to London was announced. We thanked him, bid him goodbye and were on our way. We thought.

10. We boarded Egypt Air. Felt so secure. Saw a most friendly supervisory stewardess (18 years of service), the same one from our flight from London to Cairo. She remembered us. Went

out of her way to find us the most comfortable seats. Egypt Air's stock rose. But this particular plane didn't. Something went wrong with its inner workings -- think it was the oxygen system. They emptied us back into the stand-up bus and sent us back to the terminal. Much faster than it took to empty the baggage, which seemed to take forever.

11. Plane was scheduled to take off at 9:30. We finally took off on another plane about noon. Touch and go. Without Nabil, Aleen and I were rather helpless. Couldn't understand Arabic announcements and when English was spoken, it was worse. Pins and needles for nearly two and a half hours. Every time Aleen and I would ask someone from Egypt Air when we were scheduled to take off, they gave us a ticket for free refreshments. We didn't want orange juice. We were perspiring orange juice. All we wanted was to get back home via London.

12. I should insert here that Aleen is where the action is. If not she effects action. When we were called for the first plane, she was a block away in the little girls' room. Nabil heard the Arabic announcement and I started out

KING FAISAL'S DEATH WAS A BIG SHOCK TO EVERYONE IN THIS COUNTRY, PARTICULARLY THE CRS TEAM. IT'S BEEN NEARLY 10 YEARS SINCE WE FIRST STARTED WORKING IN SAUDI ARABIA. **FAISAL** WAS A KIND OF COMMON DENOMINATOR BETWEEN **EGYPT** AND THE **U.S.** BOTH COUNTRIES LOST A **GREAT FRIEND.**

to find Aleen. I had a nice Egyptian fellow all primed to interpret when the announcement came over the speakers for the second time. He said excitedly, "It's your plane to London." Aleen was nowhere around. Finally I found her at the newsstand trying to buy some peanuts. I'm beginning to hate peanuts.

13. Passed over the Mediterranean. It's blue, sure 'nuf.

14. When we boarded this Boeing 707, there was our nice supervisor, the Egyptian stewardess, again. Greatest thing since Nabil. And what did she do? Plane wasn't a bit full so she took us up to first class. Aleen looks so much at home there. She's quality all over. As for me, I had a guilty conscience. Subconsciously, I wanted back where I belong -- with the poor.

15. Writing this while we are flying over the Alps. What a magnificent view; snow everywhere. The ridges look razor-sharp. Aleen and I think we see the Matterhorn.

16. Landed in London in good shape. Reservations at the Skyway Hotel are okay. That won-

derful guy, Jim Haecker, who is one of the most social-conscious persons I know, had this note waiting for us: "Hope your stay in Cairo, Luxor and Aswan lived up to expectations. I'll contact you in the States. Have a continued good trip and, again, thanks for a job well done. Also, I like your wife!" Watch it, Bud! ☹

5 February 75 - London to Washington to Houston

1. Now I know what cultural shock is: It's bending over looking at pretty pictures in the Royal Academy and putting your back out of whack. The flight home is miserable. I can't find a comfortable position.

2. Aleen and I went to the Royal Academy while in London to see the fantastic exhibit of Turner's oils and watercolors. And one of those freak things happened. While I was examining one of Turner's sketch books, I did something to my back. The next morning I could hardly get out of bed. Took me 15 or 20 minutes. Aleen

and I finally got hold of an English doctor who gave me some pills which helped the pain a little. Man, am I glad this thing didn't happen at the beginning of the trip.

3. While I was flat on my back at the hotel, I called Lady Ella Part to see how Sir Antony was getting along. Sure enough, it was a heart attack. He spent four days in intensive care but seems to be getting along fine now.☺ I am sorry Aleen and I were unable to see Antony and Ella. They send their regards to friends in CRS.

4. Son Bill was absolutely right when he warned us before the trip, "Don't let the PLO or the IRA get you." In Egypt we met some people from Rhodesia who thought we had a lot of guts to spend some time in London. When we went into the big store, Harrods, we were frisked at the door. You'd think the whole world is walking on eggshells.

5. Harrods was most impressive. Displayed fish on marble tables with dead pheasants as decoration. No smell. Couldn't have. They were next to the fine clothes.

6. Thinking back, what still bugs me about Egypt is what happened to Hassan Fathy's project at New Gourna across the Nile from Luxor. Reading his Architecture for the Poor was pure joy. (Read his quote on the back of the cover and see for yourself.) How sensitive he is to the physical, emotional and intellectual needs of the Arabs. When our people on the CRS team were seeking sociological/psychological bases for designing a housing project in Saudi Arabia, they found Hassan Fathy's words inspirational and meaningful. Yet New Gourna was a flop -- rejected by both the land and the people.

HERE'S OUR FRIEND, **HASSAN FATHY,** AGAIN.

7. So what good are beautiful words? You can't build with them. Or can you? I think Frank Lloyd Wright, Le Corbusier, Mies Van Der Rohe, Walter Gropius and Louis Kahn, to a certain extent, helped build with words. The CRS team has used words as bricks for building, too. Behind the design of every good building, and most certainly every great building, there are words expressing sound philosophy -- philosophy relating to structure, to climate, to function and form and to users' needs. Egypt needs a a sound architectural philosophy -- pure and

simple. It needs clear thinking as well as good brick.

8. We arrived in Houston late Wednesday night. Took a taxi home which cost nearly as much as the entire trip. Were so anxious to get back where everything worked. We walked up to our back door, only to see this note from Sally, our maid: "Don't use the commodes. Not working."

In Retrospect

I saw Egypt through architects' eyes. Biased? Of course. How else can I observe? Analyze? Evaluate? Background, training and experience have to come into play. My bias?

People are more important than bricks.

Sure, we must know about the chemistry of bricks. About properties and potentials of all building materials. About improvising with old building methods. About using new methods. New materials. New technology. That's the concern of our study group. The group consists of highly competent professionals -- Egyptians and Americans -- who know more than I about such things. Each member has a specialty. I don't know a damn thing about the chemistry of bricks. But I know how to stack them in place to form functional, beautiful buildings.

My specialty is design. I eat and sleep the stuff. I look at an object -- be it a 2000 B.C. temple, a 1975 mass-housing project or a mud brick village -- with prejudice. If it doesn't help people feel good, there's no architecture.

Buildings without architecture are big boxes; that's about all. They hurt, not help, people. Architecture concerns only those buildings having human values.

And what's this got to do with my trip to Egypt? Looking back now, I see things in their proper perspective. Here are three important facts that emerge out of my reflections and from this mess of notes:

1. <u>Egypt has no equal</u>. Ancient Egyptians of the Old Kingdom designed and built with imagination and skill. So did the Egyptians of the New Kingdom. Designers they were -- with art consciousness permeating their entire thinking. Then along came the Islamic architects, creating beautiful mosques and homes, as modern as today.

They too could design and build. They excelled in providing humanistic buildings. They were experts in natural lighting and natural ventilation. Better than any living architects, in or out of Egypt. Much better than Bill Caudill who spent years working in wind tunnels and artificial skies trying to find out how the wind and sun behave around building forms and spaces. No country has a better architectural heritage.

2. <u>Egypt has energy</u>. If not oil, its people. Egyptians are ambitious, vigorous, intellectual and friendly. Take any taxi driver. He bubbles with enthusiasm and drives like he's in the Daytona Beach Stock Car Classic. He's good. Take the competent Egyptian pros of our study group and other pros we met during our stay in Cairo. They're good. And motivated.

3. <u>Egypt has opportunities</u>. It's capable of advancing building technology. That's the underlying goal of the U.S.-Egyptian Study Group. It can be accomplished.

Yet something bugs me. It's been gnawing at me ever since we saw the mammoth housing project under construction in Suez City. The Ministry of Housing and Reconstruction, under the dynamic leadership of Osman Ahmed Osman, did the unbelievable. At Suez City, in nine months, they built 3,200 new family units, using the slowest type of construction -- hand-cut load-bearing stone walls. Record time. A great accomplishment. How can you argue with success?

When finished though, it'll be a dreary place to live. I saw similar monotonous mammoth housing projects in Russia. They're killers -- of human spirit. The Russians recognize this now and wish they had used better designers. With the same energy input, the same amount of money spent, those 3,200 units at Suez, if they had been properly designed, could have resulted in a wonderful, livable town -- functional, beautiful, inspirational.

Although Hassan Fathy's village at New Gourna was a failure, Suez City could have used some of Fathy's beautiful words, expressing clear statements of problems. Setting sociological goals. Being aware of users' needs. Those massive, indestructible stone boxes at Suez City may be a solution to the problem of instant housing, but not to the long-range problem of providing a pleasant, humanistic, lasting, livable new city.

Most certainly the project could have used the design skills of a Hassan Fathy and other highly talented Egyptians. If there are none, then Egypt should go outside for help. In fairness, I did see a few good modern buildings. So there must be a few good designers left in Egypt.

Our prime task as members of the U.S.-Egyptian Joint Study Group is to advance building technology. Our meetings during the week were fruitful. It was the consensus that 1) there should be a master plan for exploration, development, production and use of building materials; 2) Egypt needs to know what it has and what it should have in materials and methods; 3) steps should be taken to build new plants to produce critical materials; 4) existing plants should be upgraded; 5) transportation of materials from plant to site should be improved. There is a need for 6) more labor-saving devices, 7) system building and building systems, 8) better testing facilities, 9) more up-to-date technical libraries, 10) professional interchange between Americans and Egyptians. 11) Performance standards should be established for raising and guaranteeing quality of materials and construction; 12) there should be schools for technicians and skilled labor, and 13) there must be a resurgence in design thinking, training and skills.

You guessed it. It's No. 13 that interests me. That's been troubling my eyes...and my heart.

Sometime after Egypt nationalized its building industry, design got buried in the blowing sands as did some of the New Kingdom temples. Where the ancient Egyptians could design as well as build, the modern Egyptians seem able only to build. From my point of view, lack of design is the greatest weakness in Egypt's building industry.

With all the advanced technology in the world, with new ways of making bricks, with all the efficiency of construction management, and with tons of Egyptian pounds, there is no assurance that there will be good buildings: buildings that are good for people. Good buildings come about only through good design.

But good design costs money think most governmental officials in Egypt (and in our country, too). Not true. Good design never raises the price tag. Poor design is wastefully expensive.

There's no excuse for building bad buildings; it's just as cheap, and usually cheaper, to build skillfully designed buildings that serve people.

However, good design comes only from good designers. And there's definitely a designer shortage in Egypt. It's time to excavate the talent which was buried in the sands. Rediscover design. And the philosophical thinking that goes hand in hand with it. Restore Egypt's architectural heritage. Dust off the experienced, skillful designers, if they are still around. Give them a chance to help make Egypt a better place to work and live. Strive for design excellence. Encourage universities to produce new breeds of creative, talented designers in tune with today. Sensitive to human needs. Alert to Egypt's unique economy, mores and climate. If Egypt doesn't go this route, it will continue to get poorly designed projects -- buildings which are no more than dehumanized crates for storing people.

Design <u>is</u> costly, but only if it's not there.

We've used recycled stuff: Paper 100% recycled stock. Cover recycled 1937 Caudill sketch.